Canine Myxomatous Mitral Valve Disease

| Changbaig Hyun • Yong-Wei Hung |

PANMUN

Canine Myxomatous Mitral Valve Disease

ISBN: 979-11-5943-531-7 (93520)

Publication Date : November 1, 2025
Printed in South Korea

Publisher : Panmun Education Co., Ltd. Sung Kwon Liu
211, Mokdongseo-ro, Yangcheon-gu, Seoul, 07995, South Korea
http://www.medicalplus.co.kr
Managing Editor : Jai Seon Lee
Editor : Kyung Min Lee
Marketing : Ho Chul Kim, Sung Gyu Choi, Jin Hyung Kim, Myoung Han Jung, Mo Ran Kim, Ye Ryun Noh, Tae Su Han, Ye Seol Lim
Book size : 188 × 257 mm
Design : Wooilmedia Digitech

Book Introduction

CANINE MYXOMATOUS MITRAL VALVE DISEASE (CHRONIC MITRAL VALVE INSUFFICIENCY)

Canine myxomatous mitral valve disease is the definitive, all-in-one resource for understanding and managing the most prevalent acquired heart disease in dogs.

Authored by **Dr. Changbaig Hyun & Yong-Wei Hung**, a world-renowned authority on veterinary cardiology and interventional cardiac therapy, this comprehensive textbook bridges the gap between foundational cardiovascular science and cutting-edge clinical practice.

The book provides an in-depth, evidence-based review of the pathology and progression of **myxomatous mitral valve disease (MMVD)**, detailing current diagnostic methodologies, including advanced echocardiography and cardiac biomarkers. Crucially, it offers clear, systematic treatment guidelines for all stages of the disease, from asymptomatic to refractory heart failure.

With dedicated chapters on practical **case studies** and the latest **interventional**, **surgical, and pharmaceutical therapeutics**, this volume is an essential guide for veterinary practitioners, specialists, residents, and advanced students seeking to master the complexities of MMVD and significantly improve long-term outcomes for their canine patients.

Preface

Over the past several decades, my career has been dedicated to advancing the field of veterinary cardiology, with a particular focus on the profound challenges posed by **myxomatous mitral valve disease (MMVD)**. This degenerative disease, often a silent threat, remains the leading cause of heart failure in our small canine companions. It has been my enduring professional mission to contribute to better diagnostic clarity and more effective treatment options, including the development of new interventional and surgical therapies.

This textbook, *Canine Myxomatous Mitral Valve Disease*, is the culmination of my clinical experience, research findings, and years spent educating countless students and veterinarians both domestically and internationally.

My primary goal in writing this book is to provide practitioners with a single, comprehensive resource—a guide that systematically covers the fundamental anatomy, pharmacologic strategies, and the latest advancements in interventional and surgical management. I hope this volume not only deepens your understanding of MMVD's complexities but also serves as an immediately practical reference to enhance the quality of care and improve the longevity of your patients.

I extend my sincere gratitude to co-author Dr. Yong-Wei Hung, my colleagues, students, and, most importantly, the dedicated owners who entrust their beloved companions to our care.

May this book serve as a valuable tool in our shared commitment to veterinary excellence.

Dr. Changbaig Hyun, DVM, PhD

About the Author

Changbaig Hyun, AiCVIM, DVM, PhD

Dr. Changbaig Hyun obtained his **veterinary license** in **1990** and subsequently operated a small animal hospital in Gangnam, Seoul, for five years.

He was then selected as an Australian Government Scholar and completed his **Master's and Doctoral degrees** in small animal clinical studies at the **University of Queensland**, Australia. He earned his advanced degrees in the small animal clinical field (specifically **Cardiology and Genetics**) and worked as a clinical assistant at a university hospital.

Following this, he completed a **postdoctoral fellowship** at the University of Missouri College of Veterinary Medicine. In **2001**, he worked at the Victor Chang Cardiac Research Institute in Sydney, Australia. Since **2005**, he served as a Professor of Small Animal Internal Medicine at **Kangwon National University**, educating students and treating patients at the university hospital for 15 years.

Dr. Hyun is a **world-renowned authority** on **veterinary cardiology**, **interventional cardiac disease therapy**, and **hereditary heart disease**. His achievements include being listed in *the Marquis Who's Who in the World* in **2008** and receiving the **"Educator of the Year" Award** from the prestigious International Biographical Centre in the UK.

He has authored **18 books**, including treatment guidelines for small animal internal diseases and textbooks on cardiology. He has also published **193 academic papers** in globally renowned journals (e.g., *Circulation Research, American Journal of Medical Genetics, American Journal of Veterinary Research, Veterinary Journal, Veterinary Surgery, Veterinary Microbiology, Veterinary Records, and Journal of Veterinary Internal Medicine*).

In the past 15 years since returning to Korea, he has contributed to the advancement of small animal clinical practice in the country by educating numerous veterinarians through **over 300 academic lectures and training programs**.

In **2015**, he was designated as the first **Founding Specialist** in the field of small animal internal medicine (Cardiology) by the Asian College of Veterinary Internal Medicine (AiCVIM) in Korea. Due to his high international recognition, he is frequently invited by veterinary associations and academic organizations abroad (including the US, Japan, Taiwan, China, Thailand, and Malaysia) to lecture on the interventional procedures and treatments he has developed.

Yong-Wei Hung, BVM, MS, DAiCVIM, FASE

Dr. Yong-Wei Hung earned his Bachelor of Veterinary Medicine (BVM) degree from the Department of Veterinary Medicine, College of Agriculture, National Taiwan University in 1990. Following a brief period in private practice, he completed a one-year Intensive Post-graduate Training Program at the Animal Medical Center (AMC) in New York in 1991. During his time at AMC, he was profoundly influenced by Drs. Philip Fox, Si-Kuang Lu, and Robert Hamlin, which inspired him to pursue graduate studies in veterinary physiology and pharmacology at The Ohio State University (OSU) under the mentorship of Dr. Hamlin.

Upon receiving his Master of Science (M.S.) degree from OSU, Dr. Hung began a cardiology training at the North Carolina State University Veterinary Teaching Hospital. He later became a board-certified cardiologist by the Asian College of Veterinary Internal Medicine (AiCVIM). In 2024, Dr. Hung was awarded the Fellow of the American Society of Echocardiography (FASE), recognizing his contributions to the field.

After completing his formal cardiology training, Dr. Hung established Taiwan's first veterinary cardiology specialty practice. Recognizing the critical role of continued professional development, he founded the Taiwan Academy of Veterinary Internal Medicine, which later became Taiwan College of Veterinary Internal Medicine, has since grown to include over 750 active members. In 2023, he became the first Taiwanese recipient of the prestigious FASAVA–Hill's Practitioner of the Year Award.

Contents

CHAPTER 01

Basic Structure and Function of the Cardiovascular System

1.1 Basic Anatomy of the Cardiovascular System

Pericardium

The **pericardium** is a thin, tough, fibrous sac that completely surrounds the heart, containing a small amount of fluid (2.5-15 ml [0.226±0.018 ml/kg]). It consists of a fibrous outer layer and a serous inner layer. At the **cardiac base**, the pulmonary artery and aorta originating from the heart are encased within the pericardium and firmly attached to the myocardium. The pericardium is elastic and can accommodate a large volume of fluid. It can also expand continuously. However, excessive expansion can increase intrapericardial pressure, compressing the cardiac chambers and leading to heart failure (especially right-sided heart failure, as the thinner right ventricular wall is more susceptible to compression).

Lungs

In dogs, the left lung consists of a **cranial lobe** and a **caudal lobe**, with the cranial lobe further divided into an anterior and posterior part. The left cranial lobe extends beyond the first rib and is slightly craniomedial (Figure 1-1). The right lung consists of a cranial lobe, **middle lobe**, caudal lobe, and **accessory lobe** (Figure 1-1). The **pulmonary trunk** transports deoxygenated venous blood from the right ventricle to the lungs. The pulmonary trunk originates from the **pulmonary ring** located in the right ventricular outflow tract. The pulmonary veins carry oxygenated arterial blood from the lungs to the left atrium.

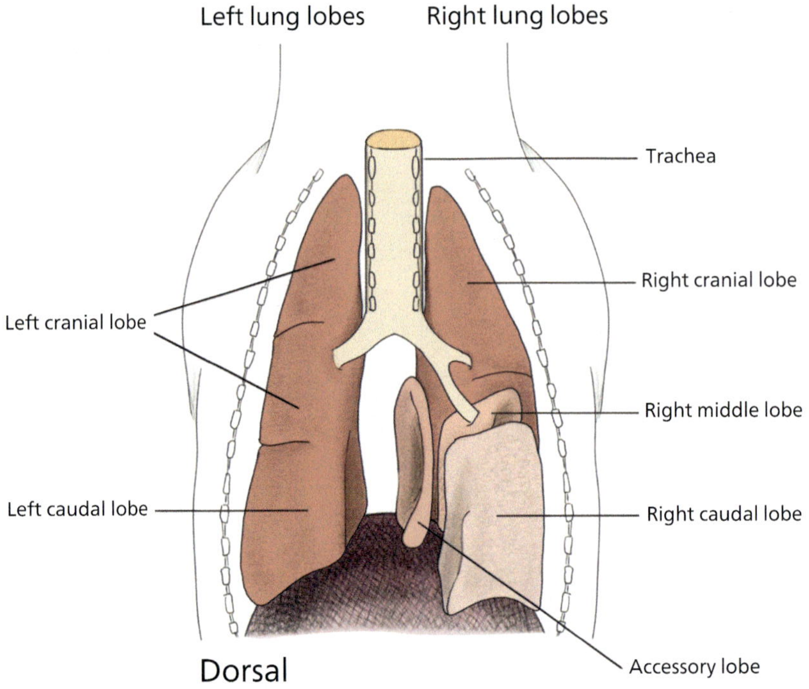

Figure 1-1 ▶ Structure of the left and right lung lobes in a dog.

Heart

The hearts of dogs and cats are large muscular pumps divided into four **chambers** by a **septum** and **valves**. The heart is primarily composed of ventricles, atria, the interatrial septum, the interventricular septum, and four valves. The right heart contains the **tricuspid valve** between the atrium and ventricle, and the **pulmonary valve** in the right ventricular outflow tract. The left heart contains the **mitral valve** between the atrium and ventricle, and the **aortic valve** in the left ventricular outflow tract. The mitral and tricuspid valves are supported by **chordae tendineae** and **papillary muscles**. The heart also consists of a muscular layer, the endocardium lining the inner surfaces of the atria and ventricles, the epicardium (a thin membranous tissue on the outside of the heart), and the pericardium that encloses the heart between the heart and the thoracic cavity.

The heart is located between the 3rd and 6th ribs, with the **cardiac base** facing dorsally and the **cardiac apex** facing ventrally. The ventricles, which pump blood out, are located caudal-ventrally, while the atria, which receive blood, are located cranial-dorsal to the ventricles.

A normal dog's heart size is approximately 0.7-0.8% of its body weight (8 g/kg). Female dogs have smaller hearts than male dogs, and smaller breeds have a larger heart-to-body weight ratio compared to larger breeds. Obese dogs may have an increased apparent heart size on radiographs due to pericardial or mediastinal fat, which can lead to a higher heart-to-thorax ratio compared to non-obese dogs. Older dogs may show changes in cardiac silhouette shape, but they do not consistently have a smaller heart ratio. In cats, the heart ratio is smaller than in dogs, approximately 0.28-0.38%. In most animals, the relative heart size (heart-to-thorax ratio) decreases with age as the thorax grows. In dogs, however, the vertebral heart score tends to increase slightly with age. Furthermore, the canine left ventricle is about 3 times heavier and thicker than the right ventricle, and in cats, it's about 3.5 times heavier and thicker. As dogs grow, both the left ventricular wall thickness and the ventricular lumen diameter increase substantially. In adults, ventricular dimensions are approximately 2–3 times larger than in puppies.

The position of the heart can be easily identified by the location of the **left auricle**, **right auricle**, and coronary arteries. The **coronary groove** surrounds most of the heart, with the exception of the pulmonary conus area which is connected to the dorsal aspect of the right ventricle. Observation of the coronary groove allows for differentiation between the atria and ventricles. When viewed from the cranial-dorsal aspect, the right auricle is located next to the **pulmonary trunk**, and the left auricle is next to the **aortic root** (Figure 1-2). The **right ventricular outflow tract (RVOT)** and **main pulmonary artery (MPA)** are most easily observed at the cranioventral aspect of the heart (Figure 1-2). The apex is formed by the left ventricle. When the heart is viewed from the caudal-dorsal aspect, the circumflex branch of the left coronary artery can be observed distributing along the left coronary groove to the left atrium and left ventricle (Figure 1-2). The left coronary artery is most clearly observed in the caudal left ventricular wall, while the right coronary artery is distributed posterior to the right auricle's right coronary groove, supplying the right ventricular wall. The **caudal vena cava** enters the right atrium, and the **cranial vena cava** enters the dorsal aspect of the right atrium (superior part of the atrium). The pulmonary veins enter the dorsal aspect of the left atrium (Figure 1-2).

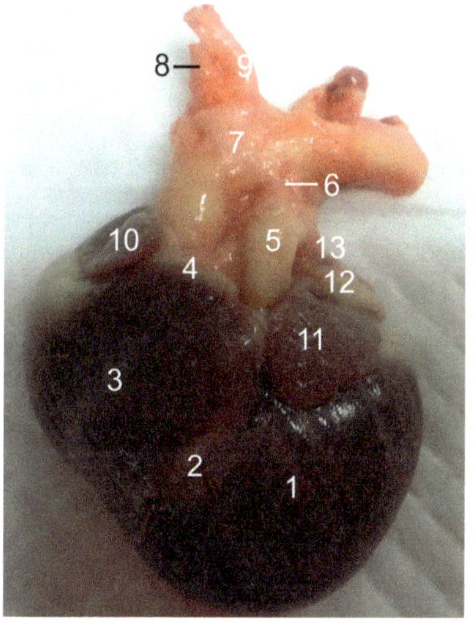

Figure 1-2 ▸ External view of the canine heart. 1. left ventricle, 2. interventricular groove, 3. right ventricle, 4. arterial cone, 5. pulmonary trunk, 6. arterial ligament, 7. aortic arch, 8. brachiocephalic trunk, 9. left subclavian artery, 10. right auricle, 11. left auricle, 12. coronary groove, 13. pulmonary veins.

The **atria** are relatively thin-walled structures. They receive blood returning to the heart from the systemic circulation (right atrium) and pulmonary circulation (left atrium), and then pump this blood into the ventricles. Therefore, the atria have a much greater capacity for expansion than the ventricles. Each atrium possesses an **auricle** (auricular appendage).

The **right atrium** is located cranial-dorsal to the right ventricle and receives blood from the systemic veins. The right auricle is a small, somewhat pendulous pouch-like part of the right atrium, adjacent to the distal surface of the pulmonary valve's right ventricular outflow tract. The following four **openings** enter the right atrium:

1. **Cranial vena cava** (receives blood from the head, neck, forelimbs, thorax, lumbar region, and adjacent posterior thoracic wall)
2. **Caudal vena cava** (receives blood from the hindlimbs, abdominal organs, and abdominal wall)
3. **Coronary sinus** (located inferior to the cranial vena cava and directly delivers blood from within the heart to the right atrium)
4. **Right atrioventricular orifice** (located between the right atrium and right ventricle, with the tricuspid valve present here)

The **fossa ovalis** is a small depression located in the middle of the right atrial wall, a remnant of a prenatal communication between the left and right atria. During necropsy, it is common to find that this opening remains patent when probed with forceps (**patent foramen ovale**).

The **left atrium** is located on the left side of the cardiac base and is connected to the pulmonary venous system (e.g., pulmonary conus and pulmonary trunk). The confluence of the veins formed three ostia at the atrial entry point that consisted of 1) right cranial and right middle pulmonary lobe veins; 2) right caudal, accessory, and left caudal pulmonary lobe veins; and 3) both the left cranial and left caudal pulmonary lobe veins of the left cranial lung lobe.. The **left atrioventricular orifice** is located between the left atrium and left ventricle, with the mitral valve present here (Figure 1-3).

The **interatrial septum** exists between the left and right atria. On the right atrial side of the septum, the fossa ovalis is present, and on the left atrial side, a flap of the **septum primum** covers the fossa ovalis (Figure 1-3).

The **ventricles** are composed of thick muscular tissue and are responsible for pumping blood into the systemic and pulmonary circulations. Within the ventricles, the **chordae tendineae** and **papillary muscles** support the mitral and tricuspid valves (Figures 1-3 and 4A). The ventricular fibers consist of a **superficial layer**, **middle layer**, and **deep layer**, with the middle layer being the thickest (Figure 1-4B).

The **right ventricle** pumps blood into the relatively low-resistance pulmonary circulation and is separated from the left ventricle by the interventricular septum. Because it can pump blood with relatively less effort than the left ventricle, its wall thickness is considerably thinner than that of the left ventricle. If the right ventricle is transversely sectioned, the ventricle appears crescent or semi-lunar shaped. The right ventricle is largely divided into

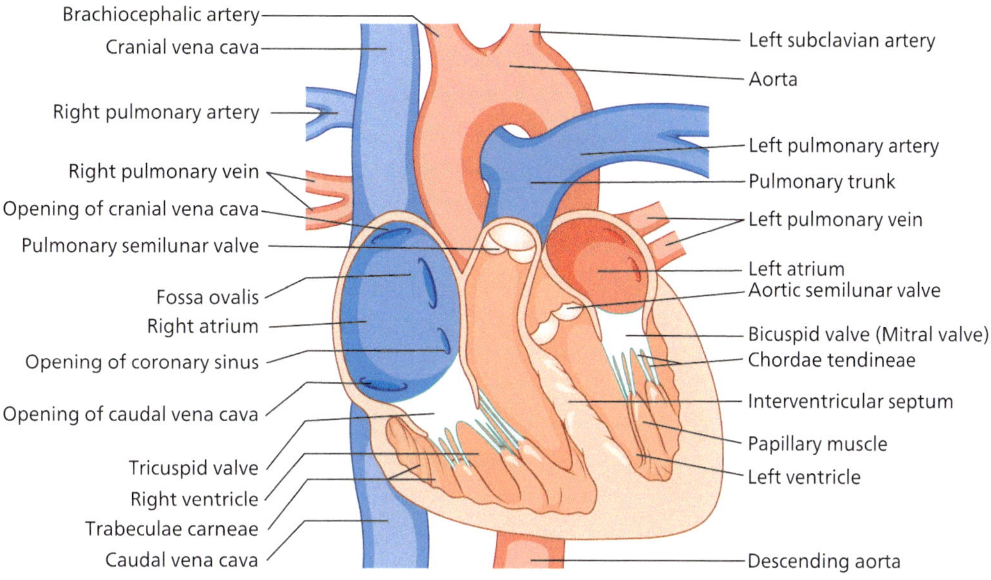

Figure 1-3 ▸ Internal structure of the heart viewed from the cranioventral aspect.

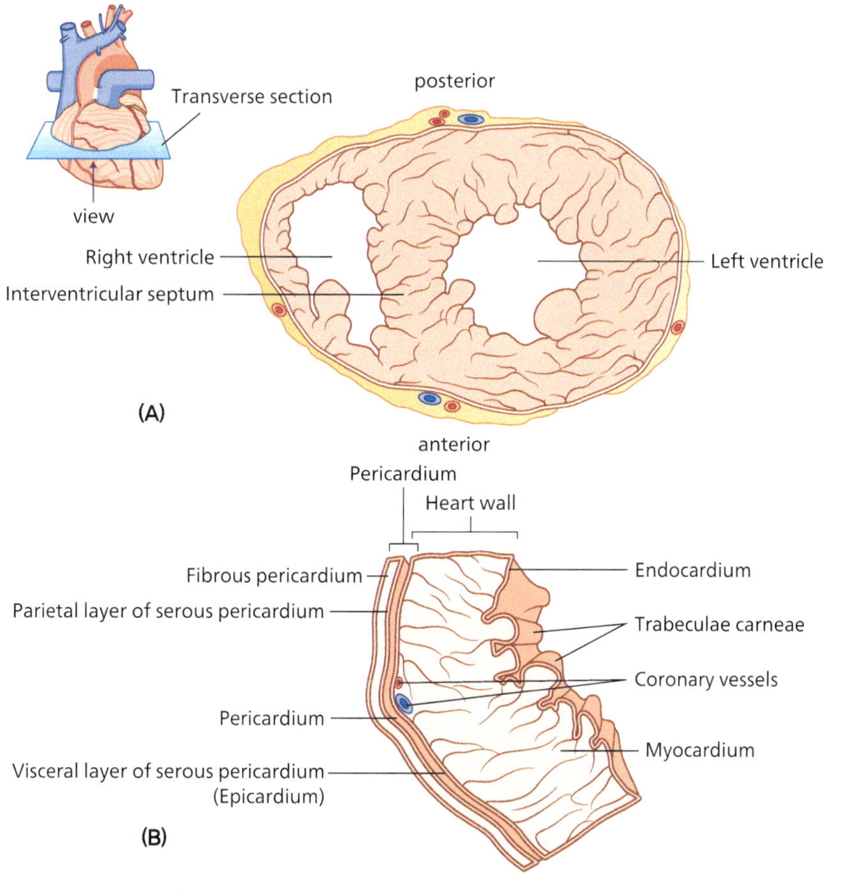

Figure 1-4 ▸ Structure of the ventricular wall. (A) Cross-sections of the heart with different thicknesses, showing varying ventricular wall thickness. (B) Structures located between the heart wall (right ventricular wall side) and the pericardium.

two parts: the **inflow region** and the **outflow region.**

The right ventricular **inflow region** refers to the **atrioventricular orifice** containing the tricuspid valve, and the right ventricular **outflow region** refers to the pulmonary orifice containing the pulmonary valve. Blood entering the right atrium passes through the tricuspid valve into the right ventricle, and when the right ventricle contracts, it pumps blood through the pulmonary artery to the lungs.

The **papillary muscles** are long, conical muscular projections originating from the right ventricular wall and connected to the chordae tendineae. Typically, three papillary muscles are observed, but their number and morphology vary due to frequent fusion (especially in cats).

The **chordae tendineae** are thin, long, tough fibrous tissues typically originating from the apex of the papillary muscles. They may also(fairly commonly seen) originate from a muscular **ventricular septal protuberance** (especially the **septal cusp** of the tricuspid valve originates here). The chordae tendineae branch into multiple secondary branches towards the valve cusps and attach to them.

Trabeculae carneae are muscular ridges, a characteristic structure of the right ventricle. These structures are especially abundant on the free wall of the right ventricle. The **trabeculae septomarginalis**, containing **Purkinje fibers** and branch bundles, connect to the ventricular wall, papillary muscles, and the free wall of the right ventricle. Refer to Figures 1-3 and 4A for the shape and location of each anatomical structure.

The **left ventricle** is a thick-walled, conical chamber that forms the apex of the heart. The anterior mitral valve leaflet separates the left ventricular inflow and outflow tracts. Blood entering through the left ventricular inflow tract is ejected through the left ventricular outflow tract, past the aortic valve, into the ascending aorta. Generally, two papillary muscles originating from the apex are present in the left ventricle: the **dorsal papillary muscle** (located slightly posteriorly) and the **ventral papillary muscle** (located slightly anteriorly). The chordae tendineae attach to the atrioventricular valve leaflets with multiple branches. One or two trabeculae septomarginalis are present, and in cats, their number and shape vary greatly.

The **interventricular septum** separates the left and right ventricles. Most of the ventricle consists of thick muscular tissue and a very thin basement membrane. Generally, the muscular layer of the interventricular septum is similar in thickness to the free wall of the left ventricle.

There are four **valves** present in the heart (Figure 1-5A). During cardiac diastole, the aortic and pulmonary valves close to prevent backflow of blood into the ventricles. Then, the atrioventricular valves (e.g., mitral and tricuspid valves) open, and blood flows from the atria into the ventricles. During cardiac systole, the atrioventricular valves (e.g., mitral and tricuspid valves) close to prevent backflow of blood into the atria. Then this blood enters the adjacent large vessels (e.g., aorta and pulmonary artery). While the atrioventricular valves are closed, the chordae tendineae stretch, causing the valves to bulge towards the atria (Figure 1-5B).

The **atrioventricular valves** are located in the **inflow tract** of the right and left ventricles and prevent blood from flowing back into the atria during ventricular systole. The atrial

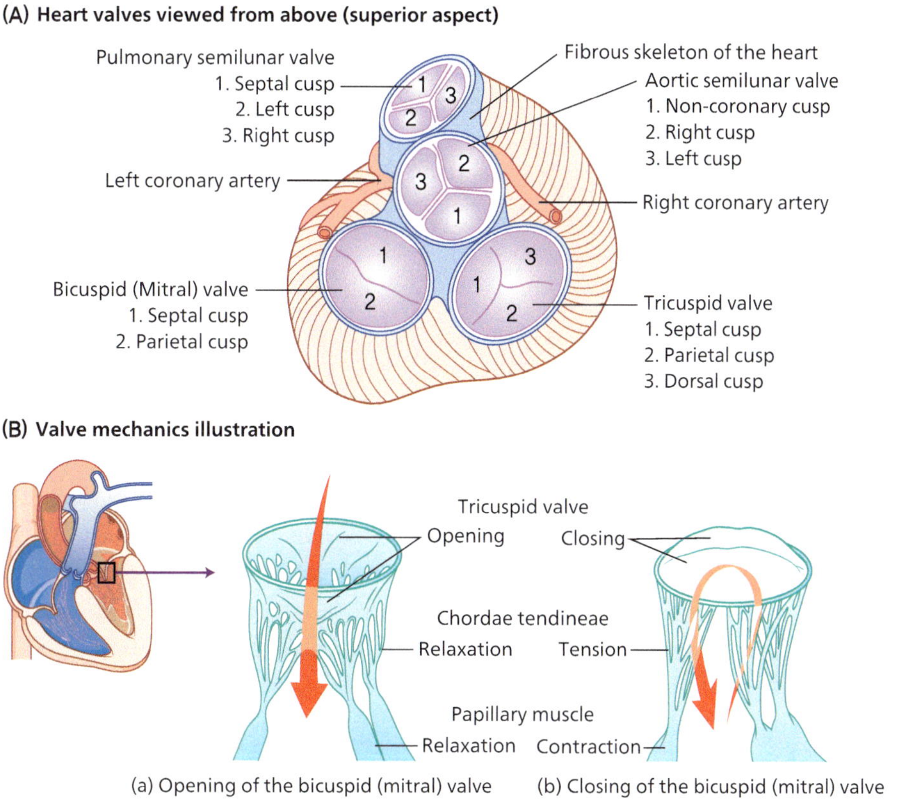

Figure 1-5 ▶ Heart valves. (A) Structure of the four heart valves. (B) Valve movement, compare the appearance of the valve when open (a) and closed (b).

side of the valve is attached by the external perimeter and the fibrous ring of the atrioventricular junction, and the chordae tendineae are attached to the ventricular side of the valve.

The **tricuspid valve** (right atrioventricular valve) has three leaflets: the **dorsal septal cusp**, **ventral cusp**, and **parietal cusp** in human. Although Alves et al. (2008) found 2, 4, or 5 cusps in 68% of the 45 dogs they examined, the valve in the dog consists basically of two cusps. The cusp of the right atrioventricular valve that attaches to the fibrous ring adjacent to the septum is called the septal cusp (cuspis septalis). The cusp that attaches to the fibrous ring adjacent to the lateral or outer wall is the parietal cusp (cuspis parietalis). The extremities of these two cusps become narrower and merge or, in some specimens, small secondary cusps are formed at these sites. These are the angular cusps (cuspis angularis). Sometimes, small secondary valve leaflets are observed at the tips of the valve (Figure 1-5A). The **mitral valve** (left atrioventricular valve) is stronger than the tricuspid valve. However, the distinction between its leaflets is somewhat less clear compared to the tricuspid valve. The mitral valve also has two leaflets: the **septal cusp-anterior cusp** and the **parietal cusp-posterior cusp** (Figure 1-5A).

The **aortic and pulmonary valves** are often called **semilunar valves**, and both the aortic and pulmonary valves are attached to fibrous rings. The aortic valve consists of right, left, and Non-coronary cusp and the pulmonary valve consists of a **right cusp**, **left cusp**, and **septal cusp** (Figure 1-5A).

At the base of the ventricles, there is fibrous tissue composed of collagen and cartilage, often called the **cardiac skeleton**. This tissue separates the atria and ventricles. The **atrioventricular bundle**, a cardiac conduction tissue originating from the atria, passes through this region and distributes into the ventricles.

Cardiac Blood Vessels

The **coronary arteries** supply blood to the myocardium. The **left coronary artery** originates from the **left sinus of aorta**. and distributes to the dorsal interventricular branch (Figure 1-6). This vessel runs somewhat obliquely, distributing to the left side of the heart. It is subdivided into left ventricular branches, right ventricular branches, and interventricular branches. The **circumflex branch** originates from the left coronary artery and distributes caudodorsally, supplying blood to the left atrium and left ventricle. The **left marginal ventricular branch** distributes along the left margin of the left ventricle. In dogs, the circumflex branch supplies blood to the posterior aspect of the left ventricle and the posterior aspect of the right ventricle. The **right coronary artery** originates from the **right sinus of aorta** and is smaller than the left. These vessels divide into very small branches, circling the right side of the coronary groove and distributing to the right auricle (Figure 1-6). These vessels supply 2/3 of the blood to the free wall of the right ventricle. Branches of the right coronary artery supply blood to the sinoatrial node (**atrial branch; sinus node artery**). Small **accessory coronary arteries** are reported in approximately 20% of dogs.

The **coronary sinus** is a large vascular dilation observed at the end of the large **coronary veins**. This part is responsible for returning blood received from the heart back into the heart. Blood that has circulated through the heart enters the right atrium through the open-

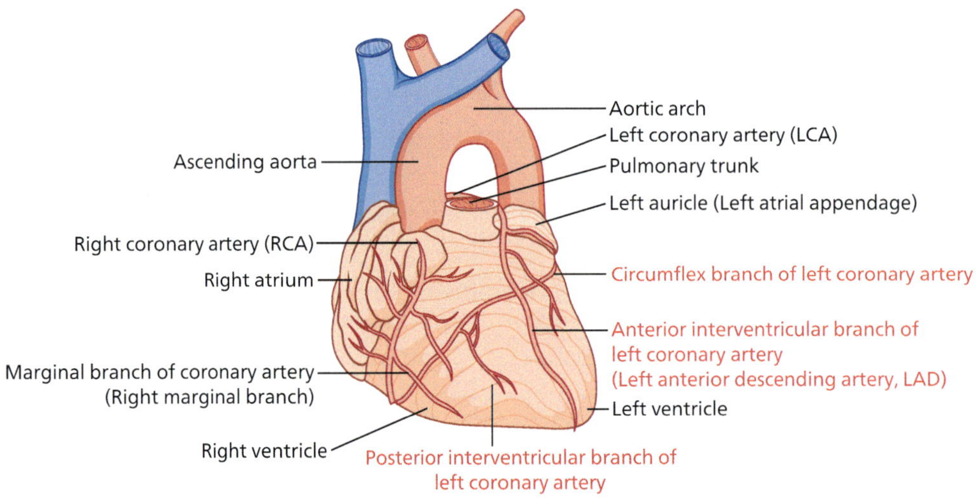

Figure 1-6 ▶ Coronary arterial system.

ing of the caudal vena cava via this coronary sinus. These vessels also have various shaped thebesian valve to prevent blood backflow.

The **pulmonary trunk** is responsible for sending venous blood to the lungs. The pulmonary trunk originates from the right ventricle, is located between the left and right auricles, and divides into the left pulmonary artery and right pulmonary artery. Proximal to the **bifurcation** where the left and right pulmonary arteries divide, there is a remnant organ called the **ligamentum arteriosum**. The ligamentum arteriosum is a ligamentous tissue formed by the regression of the **ductus arteriosus** (connection between the pulmonary artery and aorta) during embryonic development. The upper medial surface of the pulmonary trunk is in contact with the aorta, and the pulmonary artery and aorta spiral around each other. The fibrous pericardium attaches at the distal 1/4 of the pulmonary trunk.

The **aorta** is the thickest blood vessel and is responsible for sending blood ejected from the left ventricle to the systemic arteries. The ascending aorta originates from the left ventricular outflow tract and suddenly forms a curve at the caudodorsal aspect of the heart, proceeding caudally; this region is called the **aortic arch**. From the aortic arch, in humans, blood branches into three vessels (**trifurcation**) that go to the head, neck, and forelimbs. However, in dogs and cats, it branches into two vessels (**bifurcation**). The first branch, the **brachiocephalic trunk**, is the first large arterial vessel to branch from the aortic arch, running slightly obliquely to the right along the ventral aspect of the bronchus. This vessel branches the common carotid arteries and right subclavian artery. **The left subclavian artery** then branches further into smaller arteries like the vertebral artery, costocervical trunk, superficial cervical artery, and internal thoracic artery. (Figure 1-3).

Cardiac Conduction System

The heart has specialized conduction tissues consisting of the **sinoatrial (SA) node**, **atrioventricular (AV) node**, and **His bundle** (Figure 1-7). Based on the electrophysiological characteristics of the cells, the atrioventricular node can be divided into three regions: atrionodal (AN), nodal (N) and nodo-Hisian (NH). The AN region results from the convergence of the atrionodal bundles and the proximal atrioventricular bundle. This region is composed of large cells with a similar morphology to that of Purkinje cells, separated by transitional cells that have an elongated shape. These transitional cells are mixed with P cells, adipocytes, atrial myocytes, collagen and nerve fibers. The N region is composed of transitional cells closely connected with each other without interposition of connective tissue. For this reason, this region is also called compact node. The NH region is characterized by P cells and transitional cells connected with the Purkinje cells originating at the atrioventricular distal bundle.

The **sinoatrial node** is formed by thin, crescent-shaped cells and is located at the junction of the cranial vena cava, right auricle, and right atrium (Figure 1-7). The length of the node varies but generally tapers towards the caudal vena cava. Physiologically, the SA node consists of impulse-forming cells (pacemakers) that initiate cardiac conduction. Both sympathetic and parasympathetic nerve fibers innervate the SA node.

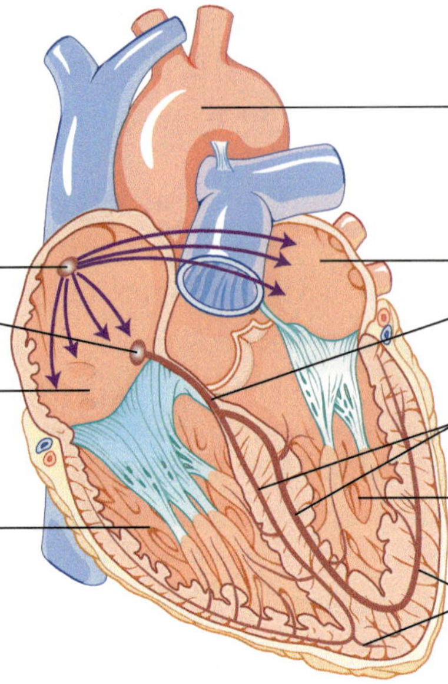

Figure 1-7 ▶ Cardiac conduction system.

The collection of cardiac conduction cells connecting the atria and ventricles is called the **atrioventricular junction**. The **atrioventricular node** is a conduction tissue composed of Purkinje cells, located directly beneath the interatrial septum on the inner surface of the right atrium (located in the **triangle of Koch**). The fibers of the atrioventricular node pass through the cardiac base and distribute downwards towards the myocardium. A series of conduction tissues called **branch bundles** are formed along the interventricular septum and the outer wall of the left ventricle. These branch bundles are divided into the **right branch bundle** and **left branch bundle** (Figure 1-7). The conduction fibers within the left ventricle are formed beneath the subendocardial tissue on the septal surface and branch into three main divisions as they descend along the septum: an **anterior branch**, a **posterior branch**, and a **septal branch** in human. In dogs, the intraventricular conduction system is trifascicular in nature. The right bundle branch is an extension of the His bundle. The left bundle branch has two divisions that fan out from the branching segment of the distal atrioventricular bundle, separately or as a common trunk. The conduction fibers of the right ventricle are spread throughout the right ventricle along the interventricular septum and the outer wall of the right ventricle (Figure 1-7).

1.2 Function of the Cardiovascular System

Cardiovascular Circulation

The normal heart of dogs and cats consists of two atria and two ventricles. Oxygenated blood exits the **left ventricle** through the aorta to the entire body, and blood rich in carbon dioxide and waste products enters the **right atrium** via the vena cava. Furthermore, this blood then enters the lungs via the pulmonary artery through the right ventricle. Gas exchange occurs in the lungs, and newly oxygenated blood re-enters the **left atrium** via the pulmonary veins (Figure 1-8).

During pregnancy, fetal blood circulation relies on the mother's placenta for oxygen supply, as the fetal lungs are atelectatic (collapsed). Additionally, this blood must remove carbon dioxide and waste products accumulated in the fetal body via the placenta. To accomplish these two tasks, embryonic vascular **shunts** (e.g., foramen ovale and ductus arteriosus) exist within the fetal heart between the right and left sides.

The entire circulatory system is broadly divided into the **systemic circulation** and the

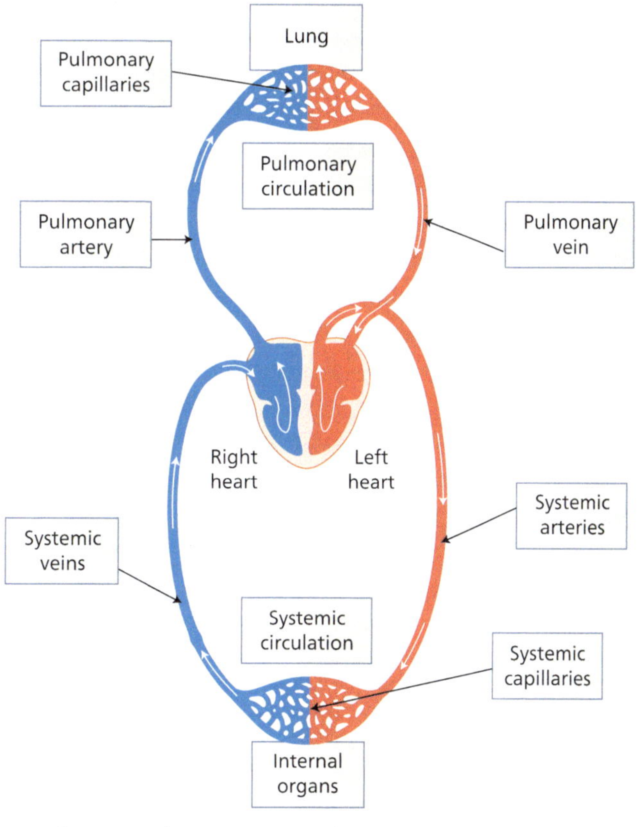

Figure 1-8 ▸ Blood circulation pathways in normal animals. Red indicates arterial blood, and blue indicates venous blood.

pulmonary circulation. These circulatory systems maintain a constant direction of blood flow through valves that open in only one direction (Figure 1-8). Oxygenated blood from the pulmonary capillaries enters the left atrium via the pulmonary veins. During left ventricular diastole, the left ventricle fills with blood, which is then ejected into the systemic circulation through the aorta during ventricular systole. The systemic circulation consists of the aorta, systemic arteries, arterioles, capillaries, and systemic veins. Large arteries have elasticity (smooth muscle), which allows them to deliver blood to peripheral tissues during cardiac diastole. Furthermore, the systemic circulation is a high-pressure system (above 100-150 mmHg), enabling blood to reach organs like the brain, which are located higher than the heart, and limb extremities. Arterioles also contain smooth muscle, allowing them to regulate **vascular resistance** through contraction and relaxation. In the **capillary bed**, oxygen is delivered to tissues, and carbon dioxide is picked up. This blood then travels through venules and veins to the vena cava and into the right atrium. During ventricular diastole, blood flows into the right ventricle, and during ventricular systole, it enters the lungs via the pulmonary artery. The pulmonary capillaries have low vascular resistance. Therefore, the blood pressure in the pulmonary circulation is much lower than in the systemic circulation (below 15 mmHg). Gas exchange occurs between the alveoli and surrounding capillaries, and the carbon dioxide exchanged with oxygen is expelled through the lungs during exhalation.

Fluid exchange between the pulmonary capillaries and interstitium is maintained by hydrostatic pressure and colloid osmotic pressure. Arterial capillaries have higher hydrostatic pressure than venous capillaries (hydrostatic pressure is higher closer to the heart). Conversely, venous capillaries have higher colloid osmotic pressure (higher with greater colloid content). Using this pressure difference, mutual exchange occurs between blood and interstitial fluid in the capillary bed (Figure 1-9).

Total blood volume is approximately 6-8% of body weight. Roughly 80% of the blood is in the systemic circulation, 15% in the pulmonary circulation, and the remaining 5% within the heart. The systemic venous system is a low-pressure, high-volume, and highly distensible system, sometimes referred to as a blood reservoir. Thus, 70-80% of blood is within the venous system. 5% of blood is in the capillaries, and the remaining 15-20% is within the arterial system. Therefore, in cases of congestive heart failure, venous vasodilators are more useful than arterial vasodilators for rapidly mobilizing blood volume.

The lungs receive blood through two pathways: 1) blood from the pulmonary artery that enters the lungs for oxygenation, and 2) oxygenated blood supplied to lung tissue by the bronchial arteries. During cardiac systole, the brain receives approximately 13% of cardiac output, the kidneys 20-25% (they receive a large amount of blood despite their small size due to their blood pressure regulation function), skeletal muscles 20%, and the digestive tract 20%. The heart, conversely, receives approximately 4% of cardiac output through the coronary arteries during diastole, not systole. During intense exercise, over 60% of blood can be concentrated in skeletal muscles.

Once blood leaves the left ventricle, it travels through the aorta, arteries, arterioles, capillaries, venules, veins, and vena cava to enter the right atrium. During this process, blood

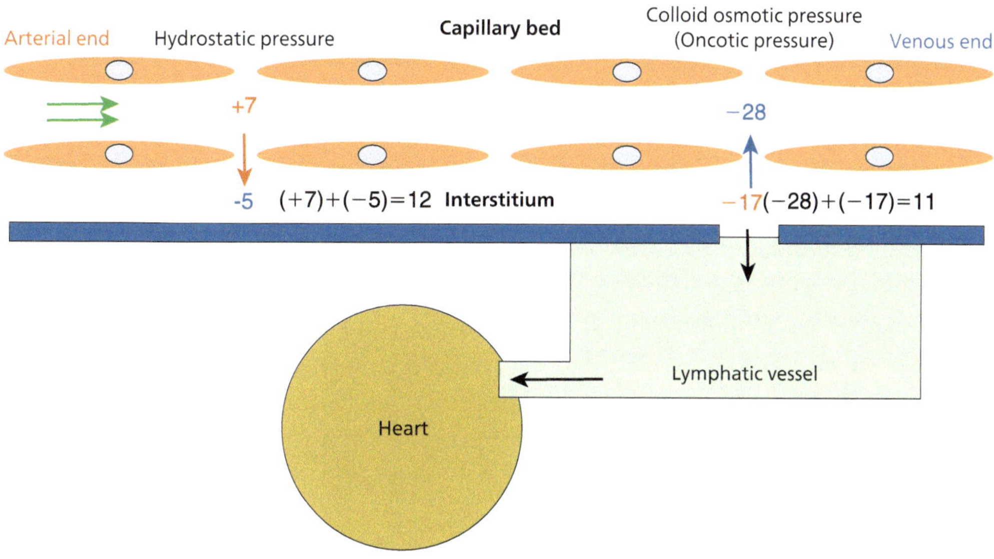

Figure 1-9 ▸ Microcirculation. Normally, the hydrostatic pressure at the arterial capillary end is much higher than the interstitial hydrostatic pressure [(+7) − (−5) = 12]. Conversely, the colloid osmotic pressure at the venous capillary end is much lower than the interstitial osmotic pressure [(−17) − (−28) = 11]. Due to this difference in hydrostatic pressure between the capillaries and interstitial fluid, oxygen and nutrients enter the interstitium, and conversely, carbon dioxide and waste products are removed due to the difference in colloid osmotic pressure. More specifically, the hydrostatic pressure difference between the capillaries and interstitium is slightly higher than the colloid osmotic pressure difference (12 − 11 = +1). Therefore, a very small amount of fluid accumulates in the interstitium. This accumulated fluid (+1) enters the lymphatic vessels and becomes lymph. Lymph enters the right atrium via the vena cava. In cases of heart failure, capillary hydrostatic pressure significantly increases [(+15) − (−5) = 20], and the amount of fluid accumulating in the interstitium greatly increases (20 − 11 = 9). However, since lymphatic vessels cannot accommodate large amounts of fluid, it is difficult to rapidly remove this suddenly increased fluid. Therefore, fluid accumulates in the interstitium, and for this reason, fluid retention in tissues, such as pulmonary edema or ascites, occurs during heart failure.

pressure drops from 100-150 mmHg in the aorta to 2-4 mmHg in the right atrium. Blood flow velocity is inversely proportional to the cross-sectional diameter of the blood vessel (blood travels quickly in the aorta/large arteries due to high flow velocity, while nutrient and gas exchange occur in capillaries due to slow flow velocity). Small arteries and arterioles account for most of the cross-sectional area of the entire arterial system and contribute approximately 50% of peripheral resistance, making them crucial for blood pressure and flow velocity regulation. Changes in arteriolar **vascular resistance** significantly impact **total peripheral resistance (TPR)**.

The **microcirculation** formed between pulmonary capillaries, interstitium, and alveoli is crucial for understanding heart disease (Figure 1-9). Fluid exchange between pulmonary capillaries and interstitium is maintained by hydrostatic pressure and colloid osmotic pressure. Hydrostatic pressure is determined by the intravascular pressure generated during cardiac output, and colloid osmotic pressure is determined by the amount of colloids, i.e., proteins, in the body fluid. Intravascular hydrostatic pressure is higher than interstitial hydrostatic pressure, and intravascular colloid osmotic pressure is lower than interstitial colloid osmotic pressure. Due to this pressure gradient, fluid (along with oxygen and carbon dioxide) exchanges between the blood vessels and interstitium. However, the difference in hydrostatic pressure between the two spaces is slightly higher than the difference in colloid osmotic pressure between the two spaces (Figure 1-9). Therefore, a very small amount of fluid accumulates in the interstitium, and this accumulated fluid re-enters the vena cava via the lymphatic vessels. However, in cases of heart failure, capillary hydrostatic pressure significantly increases, and the amount of fluid accumulated in the interstitium greatly increases. Since the rate of fluid movement into lymphatic vessels is constant, it is difficult to rapidly remove this suddenly increased fluid. Therefore, fluid accumulates in the interstitium, and this is why **pulmonary edema** occurs during heart failure (e.g., left-sided heart failure).

Cardiac Cycle

The **cardiac cycle** is broadly divided into two phases of ventricular systole (rapid ejection and reduced ejection), **isovolumic relaxation** formed between systole and diastole, two phases of ventricular diastole (rapid filling and reduced filling), **atrial systole**, and **isovolumic contraction** (Figure 1-10).

In the cardiac cycle, immediately after atrial contraction ends, all valves close, and at this time, the ventricular blood volume remains constant. Ventricular pressure rapidly prepares to rise, and this period is also called **isovolumic contraction**. When left ventricular pressure exceeds aortic pressure (pulmonary artery pressure in the case of the right heart), the semilunar aortic valve (pulmonary valve in the case of the right heart) opens. Conversely, since left ventricular pressure is higher than left atrial pressure, the mitral valve (tricuspid valve in the case of the right heart) closes. When the atrioventricular (AV) valves close, while the semilunar valves have not yet opened, the tension of the valves, together with the vibrations of the ventricular walls and the blood suddenly halted within the fixed-volume ventricles, produces the **first heart sound (S1)**. This period marks the onset of isovolumic contraction and corresponds to the formation of the R-wave, which follows the Q-wave representing ventricular depolarization on the electrocardiogram. Blood ejection from the left and right ventricle initially occurs very rapid ejection, then gradually slows down as the pressure difference between the left ventricle and aorta, and right ventricle and pulmonary artery decreases; this is called **reduced ejection**. At the end of systole, the aortic valve closes, and this closing and tensing, together with the vibrations of the aortic and pulmonary arterial walls and the blood suddenly halted within, produces the **second heart sound (S2)**. Ventricular systole refers to the period from the onset of the R-wave to the end of the T-wave on the electrocardiogram (Figure 1-10). It is also the time from when the AV valves close until the semilunar valves close. The blood volume remaining in the ventricle at this time is called the **end-systolic volume (ESV)**, and the ejected blood volume is called the **total stroke volume (TSV)**.

After blood ejection, as the pressure difference within the heart disappears, all valves

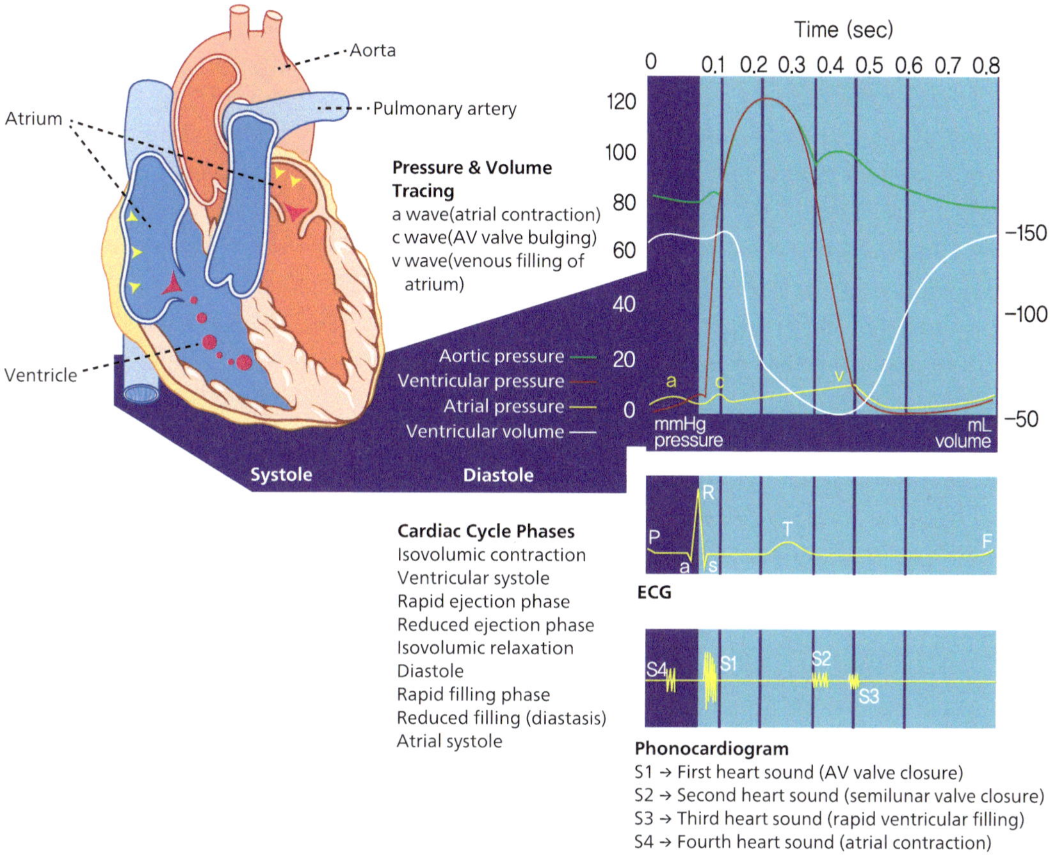

Figure 1-10 ▸ Changes in blood pressure and electrocardiogram during the cardiac cycle. The cardiac cycle consists of atrial systole - isovolumic contraction - rapid ejection - reduced ejection - isovolumic relaxation - rapid filling - reduced filling. Cardiac systole is from the R wave to the end of the T wave on the electrocardiogram. The first heart sound is formed at the onset of systole, and the second heart sound is formed at the end of systole. The third heart sound is formed during ventricular filling, and the fourth heart sound is formed during atrial systole. For video resources on the cardiac cycle, refer to the following website: http://library.med.utah.edu/kw/pharm/hyper_heart1.html

momentarily close. During this period, blood volume remains constant, and blood pressure drops. This period is called **isovolumic relaxation**. At this time, intraventricular pressure rapidly decreases. This decrease in intraventricular pressure causes the atrioventricular valves (mitral/tricuspid) to reopen, and blood from the atria rapidly flows into the ventricles. This period is called **rapid filling**, and as intraventricular pressure gradually increases, the pressure difference between the atrium and ventricle disappears. The initial period of blood flow from the atrium to the ventricle at this time is called **early diastole**, and the AV valve is maximally open. The rate of blood inflow into the atrium slows as intraventricular pressure rises; this period is called **reduced filling** (Figure 1-10). In animals with very fast heart rates, these two periods may not be distinguishable. If the ventricular wall is thin and stiff at this time, rapid deceleration of blood associated with the end of the rapid filling phase can result in a transient, low-frequency **third heart sound (S3)**. As intraventricular pressure rises again, the AV valve closes. This is represented from the end of the T-wave to the onset of the P-wave on the electrocardiogram. The aforementioned filling periods are passive blood filling, and the ventricles receive blood once again (accounting for approximately 15-20% of the filling volume) due to atrial contraction. At this time, the mitral valve rapidly opens and closes again. Therefore, the movement of the mitral valve on M-mode ultrasound appears as an 'M' shape (the movement of the mitral valve proceeds from full open → partial closure → open → full closure, hence the 'M' shape; conversely, the aortic valve movement proceeds from open → closure, thus appearing as '-o-'; refer to the cardiac ultrasound section). The **fourth heart sound (S4)** occurs during a phase of the cardiac cycle when the ventricle is expected to rapidly receive a large volume of blood—specifically, immediately after atrial contraction and just before ventricular systole. The mechanism is caused by an abnormally thick and stiff ventricle, abruptly decelerates the inflowing blood, generating audible low-frequency vibrations during this normally rapid filling phase. S4 is also referred to as an atrial gallop or a presystolic gallop. At this time, the maximum blood volume in the ventricle is called the **end-diastolic volume (EDV)**. Cardiac diastole refers to the period from the end of the T-wave to the beginning of the QRS wave on the electrocardiogram. The value obtained by subtracting the **end-systolic volume (ESV)** from the **end-diastolic volume (EDV)** is called the **stroke volume** or **total stroke volume**.

Cardiac Contraction and Relaxation

Ventricular stroke volume (or performance) is primarily determined by three interacting factors: myocardial contractility, preload, and afterload. Contractility refers to the intrinsic **inotropic strength** and pumping ability of the myocardium, independent of loading conditions. **Preload** is the degree of ventricular myocardial fiber stretch at the end of diastole, corresponding to the end-diastolic volume (EDV) or pressure. **Afterload** is the force opposing myocardial fiber shortening during ventricular systole, which is the pressure the ventricle must overcome to eject blood into the aorta. Afterload is sometimes incorrectly simplified as total **peripheral resistance** (TPR).

The myocardium is composed of millions of contractile proteins called **sarcomeres**, consisting of **actin** (thin filaments) and **myosin** (thick filaments), as well as various **cytoskeletal proteins** such as troponin and tropomyosin, which are involved in regulating this contractile movement (Figure 1-11). Myocardial contraction is initiated and sustained by ATP hydrolysis and calcium influx. Although the exact mechanism of myocardial contraction has not yet been fully elucidated, the most widely accepted theory is the **cross-bridge cycle theory**, where myosin attached to the Z-line performs a hinge movement. According to this theory, tropomyosin exists between actin and myosin, preventing cross-bridge formation between actin and myosin. Additionally, the myosin head has a heavy meromyosin head capable of hydrolyzing ATP. When calcium influx occurs and binds to troponin C connected to tropomyosin, the tropomyosin between actin and myosin detaches, and the

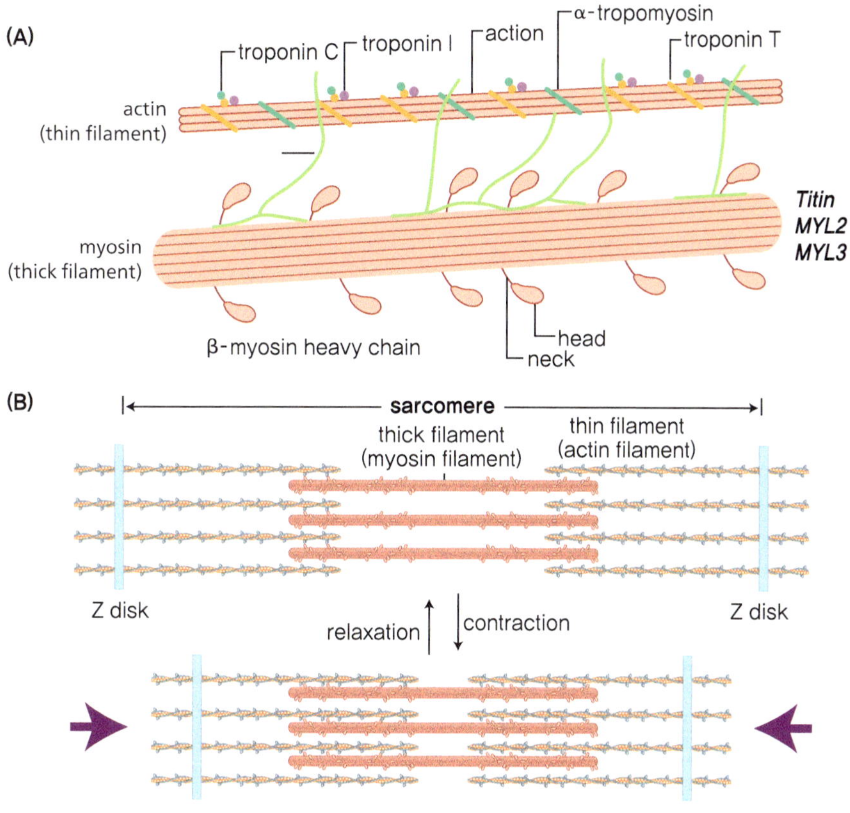

Figure 1-11 ▸ Structure of the sarcomere, the myocardial contractile unit. (A) The myocardium consists of millions of myocardial contractile proteins called sarcomeres, composed of actin (thin filaments) and myosin (thick filaments), as well as various cytoskeletal proteins such as troponin and tropomyosin, which are involved in regulating this contractile movement. (B) Changes in the sarcomere during contraction and relaxation.

myosin head begins to hydrolyze ATP, leading to cross-bridge formation and the hinge movement of myosin (myocardial contraction). As the intracellular calcium concentration increases, more cross-bridges are formed, and a stronger force required for contraction is exerted. Myocardial relaxation, conversely, occurs as calcium concentration decreases, leading to the loss of cross-bridge activity (Figure 1-11).

Intracellular calcium concentration in myocardial cells is regulated by the **Na/Ca exchanger** on the cell membrane, the **plasma membrane Ca pump**, and the **sarcoplasmic reticulum (SR)** within the cell. If depolarization of myocardial cells occurs due to an increase in sodium ion conductance, calcium flows into the cell, and calcium stored in the SR is released. This calcium binds to tropomyosin, initiating myocardial contraction. Beta-adrenergic stimulation increases intracellular calcium concentration, leading to the formation of more cross-bridges and thus stronger contractility. It also activates ATPase to produce more energy. Therefore, **beta-blockers** are sometimes used in patients with myocardial hypertrophy to weaken contractility and reduce heart rate.

Frank-Starling Law: If all other conditions are equal, the greater the blood volume within the ventricle (e.g., the stronger the **preload**), the stronger the contractile force exerted. This principle is known as the Frank-Starling Law. This phenomenon occurs because stretched myocardial fibers sensitize contractile proteins, leading to a higher increase in calcium concentration, and thus more cross-bridges are formed, resulting in greater tension.

Myocardial relaxation occurs as calcium bound to troponin C moves out of the cell or returns to the SR, causing the hindrance of tropomyosin between actin and myosin to reappear. The ability of the myocardium to relax is an important factor in determining ventricular filling pressure. The main factors determining this relaxation ability include **ventricular compliance** or **ventricular stiffness**. The myocardium becomes stiff when filled with blood. Therefore, if right atrial pressure increases, leading to increased blood volume in the coronary circulation, the ventricle becomes stiff. Additionally, if fibrous tissue infiltrates the ventricle, the elasticity of the myocardium is lost, leading to stiffness. Various factors are involved in this myocardial relaxation ability. The most important is how much calcium ions dissociate from troponin C. Furthermore, since ATP hydrolysis is the energy source for myocardial contraction, the oxygen saturation status (e.g., presence or absence of hypoxia) is also an important variable. The heart's ability to expand is equal to the ventricular external pressure (e.g., intrathoracic pressure) minus the intraventricular pressure (e.g., end-diastolic pressure). This ability also depends on ventricular blood volume (preload). In other words, the greater the ventricular blood volume, the stiffer and less compliant the ventricle becomes, requiring more **transmural pressure** to stretch. Clinically, diseases that impede ventricular filling mechanisms include **hypertrophic cardiomyopathy**, **myocardial fibrosis**, **cardiac tamponade**, **myocardial infarction**, and conditions that increase intrathoracic pressure.

Cardiovascular Regulation

Cardiovascular regulation occurs to: 1) simultaneously supply optimal blood to all organs, 2) maintain optimal cardiac activity and blood pressure, and 3) safely redistribute blood volume from inactive organs to active ones. This regulation of the cardiovascular system occurs by adjusting the tension of vascular smooth muscle, thereby inducing changes in blood vessel diameter. Vascular tension is maintained by: 1) autoregulation, 2) neural regulation, and 3) hormonal regulation.

Central nervous system control of circulation occurs in the medulla oblongata. The pathways of these central receptors are distributed from high-pressure areas (aorta and carotid arteries) to low-pressure areas (vena cava and atria) and to the left ventricle. These receptors primarily respond to atrial **stretch**. This induces a homeostatic response, where stimuli generated during systole are transmitted to the heart and blood vessels, regulating blood pressure and heart rate. **Baroreceptors** are distributed in the **carotid sinus** and **aortic arch**, sensing changes in systemic arterial blood pressure. Baroreceptors in the carotid sinus send signals about blood pressure changes to the medulla via the glossopharyngeal nerve, while baroreceptors in the aortic arch communicate blood pressure changes to the medulla

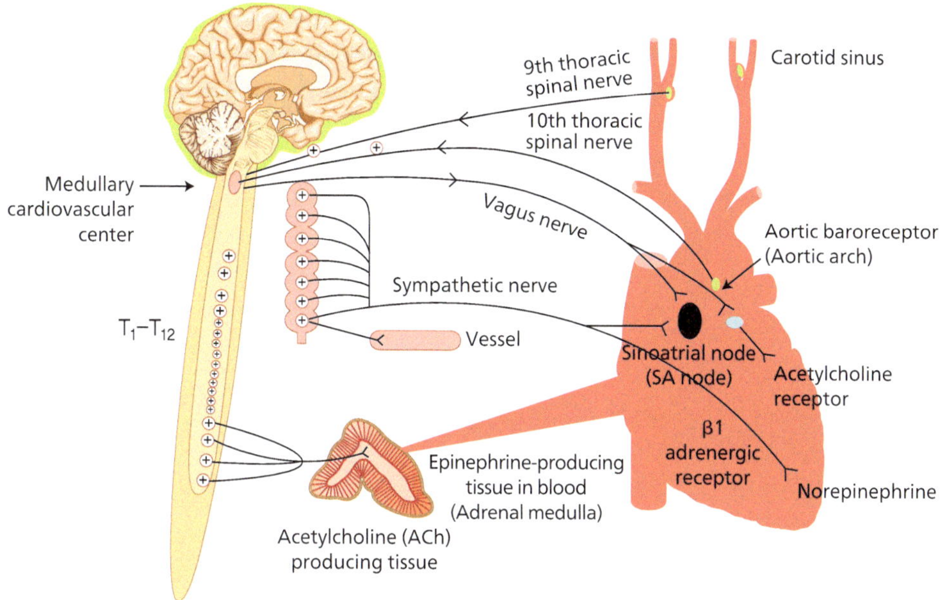

Figure 1-12 ▸ The autonomic regulation of the heart involves both the **sympathetic** and **parasympathetic (vagal)** nervous systems. **Sympathetic nerves** release norepinephrine (NE), stimulating β1 receptors, which increases heart rate and contractility. **Parasympathetic (vagus) nerves** release acetylcholine (ACh), binding to muscarinic receptors, slowing the heart rate. Reflex input from the carotid sinus and aortic baroreceptors influences the medullary cardiovascular center to adjust autonomic output accordingly. The **adrenal medulla** releases epinephrine into the bloodstream, further enhancing cardiac activity.

via the vagus nerve (Figure 1-12).

The **autonomic nervous system** that regulates the heart consists of the sympathetic nerves originating from the thoracolumbar spinal cord and the parasympathetic nerves originating from the **craniosacral nerves**. When blood pressure falls, sympathetic nerves are excited, and parasympathetic nerves are inhibited. **Norepinephrine (NE)** is released from sympathetic nerve endings, and this NE stimulates alpha-adrenergic and beta-adrenergic receptors. Alpha1-adrenergic receptors are distributed in vascular smooth muscle, causing vasoconstriction (especially in veins). Beta2-adrenergic receptors are also distributed in vascular smooth muscle, and if alpha1-adrenergic receptors are blocked, beta2-adrenergic receptors induce vasodilation. Beta2-adrenergic receptors are primarily distributed in the **sinus node** of the atrium, and some are also in the ventricles. The **chronotropic** effect (increasing heart rate) of beta2-adrenergic receptors is stronger than their **inotropic** effect (strengthening ventricular contractility). Conversely, beta1-adrenergic receptors have a strong inotropic effect. In addition, the existence of beta3 and beta4-adrenergic receptors has been reported, but research is still ongoing. This sympathetic excitation leads to: 1) increased heart rate, 2) strengthened myocardial contractility, and 3) increased **systemic vascular resistance** (vasoconstriction). The effect of sympathetic nerves on the venous system is to increase **venous return** by constricting venous blood vessels. It also promotes renin secretion in the kidneys, which in turn promotes sodium and water reabsorption. All these actions ultimately contribute to increasing blood pressure.

The **parasympathetic nervous system** is more widely distributed in the atria than in the ventric, and it releases **acetylcholine (ACh)** from nerve endings to stimulate muscarinic receptors. In the sinus node, it slows the rate of depolarization and causes vasodilation by relaxing vascular smooth muscle. In heart rate regulation, the influence of the vagus nerve (parasympathetic nerve) is greater than that of the sympathetic nervous system. The parasympathetic nervous system has opposing effects to the sympathetic nervous system and plays a role in lowering blood pressure.

Various hormones are involved in maintaining blood pressure and regulating vasoconstriction. Prominent examples include **Angiotensin II (AGII)**, **antidiuretic hormone (ADH)**, and **endothelins (ETs)**. These hormones respond to hypovolemia and decreased blood flow by inducing vasoconstriction, thereby increasing **systemic vascular resistance (SVR)** and raising blood pressure. AGII is formed by renin and increases blood pressure through vasoconstriction (Figure 1-13). It also increases sympathetic activity in the glomerulus. AGII has two receptors (AT1 and AT2). AT1 is primarily distributed in blood vessels, and AT2 is primarily distributed in the heart. These receptors cause AGII to constrict vascular smooth muscle, promote NE release to excite sympathetic nerves, and specifically, AT1 receptors distributed in the myocardium have an effect that promotes myocardial hypertrophy.

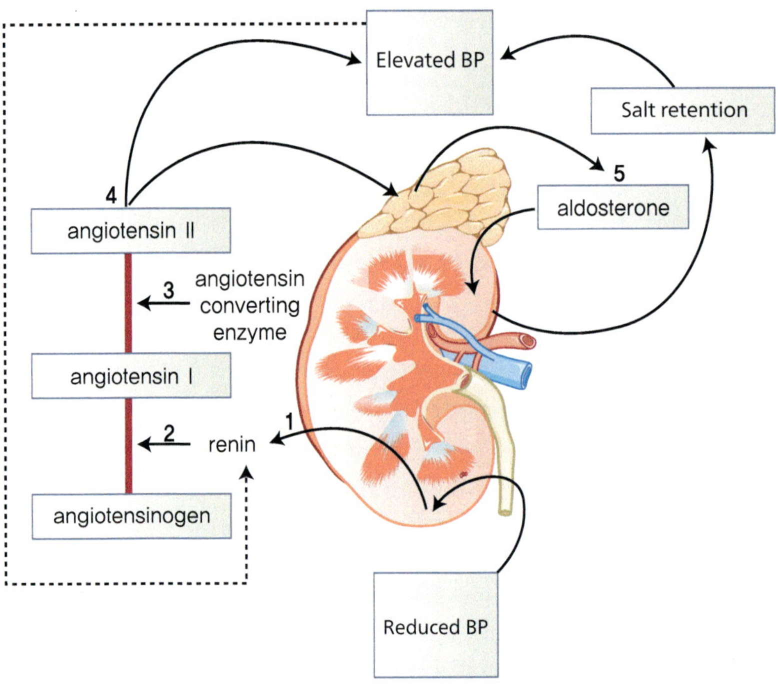

Figure 1-13 ▶ Renin-angiotensin-aldosterone system (RAAS). When blood pressure drops, the kidneys release renin, and renin ultimately releases angiotensin II (AGII) and aldosterone, which retain salt and constrict blood vessels, thereby raising blood pressure.

ADH is secreted from the posterior pituitary gland and has two receptors in the body (V1 and V2). V1 causes vasoconstriction, which increases blood pressure (though this effect is stronger with AGII). V2 has a water-retaining effect in the kidneys.

ETs are potent vasoconstrictors, and three endothelins have been discovered (ET1, ET2, ET3). There are two receptors in blood vessels (type A, type B). Type A receptors have a particularly high affinity for ET1 rather than ET3. Conversely, type B receptors have an equal affinity for all endothelins. Type A has a powerful vasoconstrictive effect. NE and AGII promote the secretion of these ETs. In cases of heart failure in dogs and humans, circulating ETs concentrations are reported to be elevated. ETs, similar to AGII, increase heart rate, increase myocardial contractility, and increase blood pressure. They also induce myocardial hypertrophy.

Venous return is a factor that determines ventricular filling and cardiac output. Blood volume regulation is crucial for regulating arterial pressure and blood flow velocity. Blood volume also significantly impacts venous pressure and capillary pressure. For example, in severe heart failure, if ventricular blood volume increases by approximately 25-30%, end-diastolic ventricular pressure increases, leading to a concomitant rise in venous and capillary pressures, causing **fluid retention** (e.g., pulmonary edema, ascites).

Changes in blood volume within the cardiac chambers are detected by **mechanoreceptors** distributed in the atrial and ventricular walls, the great vein junctions, the pulmonary artery, and the subendocardial coronary vasculature. The detected signals are transmitted through the sympathetic and parasympathetic nerves in the medulla, inducing changes in vascular resistance and heart rate, allowing for flexible responses to changes in blood volume within the cardiac chambers. The vagus nerves distributed within the atria and ventricles detect changes in cardiac chamber volume and myocardial contractility, inhibiting sympathetic nerves and exciting parasympathetic nerves.

The **kidney** is the most important organ in blood volume regulation because it controls the reabsorption of sodium and water. If blood volume or cardiac output decreases, renal blood flow decreases. This leads to an increase in afferent arteriolar pressure and excites the sympathetic nervous system. It also reduces sodium ion influx in the macula densa. These responses promote the secretion of **renin**. Ultimately, increased renin secretion increases the synthesis of AGII and aldosterone in the renal cortex. This hormonal secretion system is called the **renin-angiotensin-aldosterone system (RAAS)** (Figure 1-13). Renin secretion is regulated by pressure centers in the aorta, sympathetic nerves, changes in the composition of tubular fluid reaching the macula densa, and prostaglandins. **Aldosterone** promotes the excretion of potassium and hydrogen ions and promotes the reabsorption of sodium, thereby increasing the influx of water into the blood and ultimately increasing blood volume. AGII increases thirst and increases the desire for salt intake, thus indirectly retaining sodium and increasing blood volume. It also directly acts on the proximal tubules to promote sodium reabsorption. Furthermore, it constricts renal arterioles, reducing renal blood flow and thus decreasing the **glomerular filtration rate (GFR)**. A decrease in GFR leads to an increase in plasma urea nitrogen (BUN) and creatinine concentrations. Although AGII constricts all renal arterioles, the constriction of efferent arterioles is stronger than that of afferent arterioles, resulting in an increase in both blood volume and blood pressure. Sodium excretion and retention play a crucial role in maintaining blood volume. In particular, sodium ion excretion is largely dependent on a third factor not mentioned previously: GFR. **Atrial natriuretic factor (ANF)**, also released by **atrial stretch** and tachycardia, also plays an important role in maintaining blood volume. ANF binds to receptors in the nephron, dilating afferent arterioles and constricting efferent arterioles, thereby increasing renal GFR. This inhibits sodium ion reabsorption in the collecting ducts. Furthermore, it inhibits the synthesis of renin and angiotensins, and also inhibits aldosterone and ADH secretion. However, ANF is not a potent hormone. Therefore, studies in humans show that even increasing ANF concentrations in patients with end-stage heart failure cannot reverse the sodium and water retention phenomena caused by other factors.

When blood pressure decreases, atrial pressure receptors stimulate the posterior pituitary gland to promote the secretion of **vasopressin (ADH)**. This response causes thirst and retains water in the renal collecting ducts. Osmoreceptors can also induce vasopressin secretion but are not as strong as pressure receptors. Vasopressin secretion is inhibited by atrial **stretch** and enhanced by decreased atrial blood volume.

While much research has been conducted on blood pressure and blood volume, relatively little research has been done on the determinants of blood flow velocity. Blood flow velocity is regulated by three important factors: 1) **systemic arterial pressure**, 2) **venous and capillary pressure**, and 3) **venous oxygen partial pressure**. One study reported that the pulmonary artery partial pressure of all mammals is consistently maintained at ~40 mmHg, suggesting that venous oxygen partial pressure plays a very important role in determining blood flow velocity. Additionally, chemoreceptors distributed in the carotid body and aortic arch detect changes in oxygen partial pressure and regulate the rate of respiration. Myocytes distributed in the lungs have HEME-linked proteins (similar in nature to proteins that regulate erythropoietin secretion in the kidneys), and these proteins act as O_2 sensors. In hypoxic conditions, they constrict blood vessels, increasing blood flow velocity.

Since blood pressure is a major determinant of blood flow velocity, **impedance** to blood flow and **systemic vascular resistance (SVR)** become key determinants of blood flow velocity. For example, aortic blood flow velocity is determined by: 1) aortic diameter, 2) aortic wall elasticity, 3) SVR formed in the peripheral arterial system, and 4) extrinsic factors such as blood cell count and blood **viscosity**. Ultimately, blood flow velocity should be considered to be determined by factors affecting **afterload**. Blood viscosity is determined by **PCV (packed cell volume)**, serum protein concentration, and the **foldability** of red blood cells (the resistance encountered when large red blood cells pass through small capillaries).

Cardiac Conduction Pathway

The most important mechanism of cardiac conduction cells is to maintain regular and systematic cardiac contraction. This cardiac contraction is triggered by electrical impulses called **action potentials**. Cardiac cells have a thin membrane, and various ions (carrying positive or negative charges) move across the cell membrane through specific ion channels, generating electrical charges. These channels selectively allow only certain ions to pass, and therefore, ion concentration differences occur depending on changes in the **permeability** of these ions within the cell membrane. This results in different charges within the cell membrane (depending on the ion concentration difference), which is called the **membrane potential**. Normal cells usually have a slightly negative charge inside, and in the case of cardiac cells, they have a charge of -70 to -80 mV, which is called the **resting membrane potential**. When a cell is excited, this negative charge rapidly changes to a positive charge, which is called an action potential, and this causes the myocardial cell to contract. This action potential is generated because various ions move across the cell membrane, causing changes in membrane permeability.

A. **Cardiac Pacemaker** The myocardium is broadly composed of conduction cells that transmit electrical impulses to the heart and myocardial cells that cause contraction. Typically, the electrical impulses that control cardiac contraction and relaxation originate from the **sinoatrial node (SA node)**. However, these pace-making electrical impulses occur throughout all heart tissues. Only the impulses from the SA node occur fastest, which is why the SA node acts as the heart's pacemaker. Electrical impulses are conducted from the SA node to the **atrioventricular node (AV node)**, the **His bundle**, the **right and left branch bundles**, and **Purkinje cells** to the ventricles. If the rate of action potential generation in other conduction cells increases due to ischemic damage, inflammation, or necrosis, or if the rate of action potential generation in the sinus rhythm slows, pacemaking can occur in areas other than the SA node, leading to arrhythmias.

B. **Depolarization** refers to the period when the intracellular charge rapidly changes from negative to positive, and **repolarization** refers to the period when the intracellular charge returns from positive to negative. These changes are caused by the movement of Na^+, K^+, and Ca^{2+} ions (depending on the opening and closing of ion channels). The types of ion channels involved in the generation and extinction of action potentials vary somewhat among the SA node, AV node, atrial cells, and Purkinje cells. For example, the ion channels involved in ventricular action potential generation and extinction are as follows (Figure 1-14):

- **Phase 0:** Phase 0 is the period of rapid depolarization when the negative resting charge changes to positive due to increased Na^+ ion permeability and decreased K^+ ion permeability.
- **Phase 1:** Phase 1 is the period of early repolarization for a short duration, primarily involving K^+ and Cl^- ions (I_{to} current). This can be seen in the action potential of Purkinje cells but not in the action potential of pacemaker SA nodes. This early repolarization is called **action potential overshoot**.
- **Phase 2 & 3:** Phase 2 is the period that maintains a **plateau**, balancing the influx of Ca^{2+} ions and the efflux of K^+ ions. This phase lasts for a long time, especially in myocardial cells, and gradually transitions to phase 3 as other potassium channels

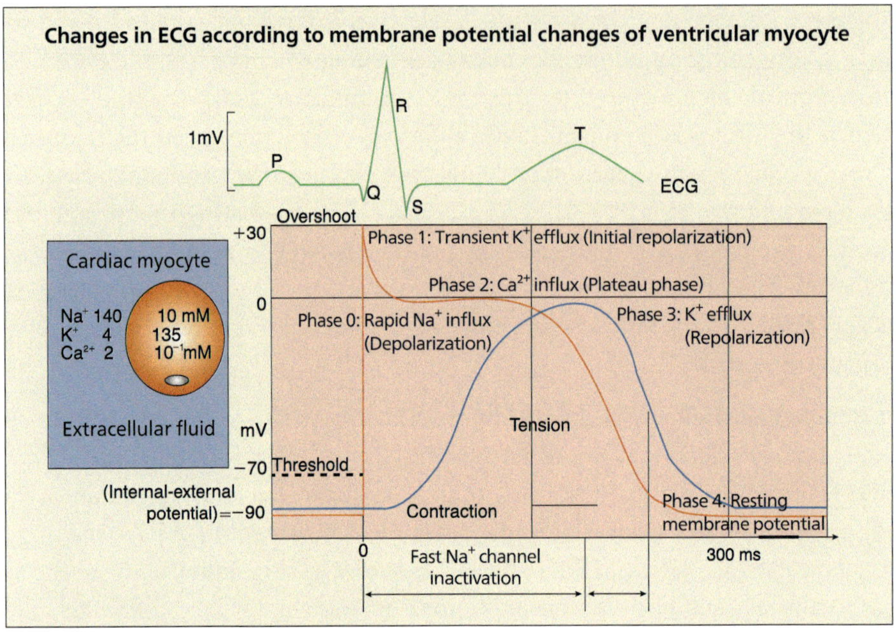

Phase 0: Rapid Na^+ influx (Depolarization)
Phase 1: Transient K^+ efflux (Initial repolarization)
Phase 2: Ca^{2+} influx (Plateau phase)
Phase 3: K^+ efflux (Repolarization)
Phase 4: Resting membrane potential

Figure 1-14 ▸ Formation of ventricular action potential and electrocardiogram changes due to ion channel changes (refer to text). The ventricular action potential consists of 5 phases (phases 0-4). Phase 0 is the period of rapid Na^+ (I_{Na}) influx, causing depolarization of ventricular cells. Phase 1 is the early repolarization phase formed when sodium channels close and potassium channels (transient outward K^+ currents; Ito) open. Phase 2 is the plateau state where calcium channels (inward Ca^{2+} currents; I_{CaL}) and potassium channels (outward K^+ currents; I_{Kr}, I_{Ks}) open, balancing the charges. Phase 3 is the period when various potassium channels (outward K^+ currents, including I_{Kr} & I_{Ks}) open, entering the final stage of repolarization. Phase 4 refers to the period when another potassium channel, I_{K1}, opens, ending repolarization. The resting potential is −80 mV, and as potassium flows in, the voltage gradually increases, and if it crosses a certain threshold, sodium channels momentarily open, causing depolarization. At this time, the QRS wave is initiated on the electrocardiogram.

open. During this period, repolarization occurs, where the intracellular charge returns to negative.

- **Phase 4:** Phase 4 is when the charge returns to the resting negative charge. This state then increases sodium influx again, triggering another action potential, hence it is also called **pacemaker potential** or **diastolic potential.**

The action potential described so far is mainly seen in non-pacemaker cells like Purkinje cells, and the action potential of cells like the SA node is somewhat different. In these cells, action potential depolarization is triggered by the rapid opening of rapidly inflowing calcium channels, and phases 1 and 2 are minimal. Various ion channels are involved in the initiation and maintenance of these action potentials, and the opening and closing of these channels primarily depend on the intracellular membrane potential. Furthermore, these channels have **activation**, **inactivation**, and **resting** states, and inactivation and resting states are distinct. Antiarrhythmic drugs primarily act selectively on activated/inactivated channels and do not affect resting channels. **Membrane responsiveness** is a term that describes how quickly the cell membrane responds to ion exchange within the cell. If responsiveness is diminished, the duration of the action potential is prolonged (because ion channel responses are slower, leading to slower ion changes within the cell membrane). **Refractoriness** refers to the period during which a cardiac cell recovers to its normal resting membrane potential after depolarization.

CHAPTER 02

Therapeutic Strategies and Medications for Heart Failure in Small Animals

Treating heart failure (HF) primarily aims to:

1. **Correct the underlying cause** of the heart failure.
2. **Improve secondary complications** arising from HF.
3. **Reduce the heart's workload** to enhance systemic hemodynamic status.

While advancements in veterinary medicine allow for non-surgical interventions for conditions like patent ductus arteriosus, valvular stenosis, and septal defects, and heartworm disease is treatable with immiticide®, the exact causes and definitive cures for myocardial and degenerative valvular diseases in dogs and cats remain elusive. Consequently, treatment for most small animal heart diseases still relies heavily on **supportive care**.

Before initiating supportive therapy, a precise evaluation of the patient's HF status is crucial, as the clear benefits of prophylactic medications or dietary supplements in veterinary medicine are often unproven, unlike in human medicine.

Goals of Heart Failure Supportive Therapy:
The primary goals of supportive therapy for heart failure include:

1. **Alleviating fluid retention** (e.g., pulmonary edema, ascites).
2. **Improving cardiac output (CO)**.
3. **Reducing the heart's workload**.
4. **Treating and preventing arrhythmias**.
5. **Inhibiting thrombus formation**.
6. **Minimizing damage to other organs** (e.g., kidneys, pancreas).

To achieve these goals, various medications are commonly used in small animal practice:

- **Diuretics:** Reduce preload to control fluid retention.
- **Inotropes:** Enhance myocardial contractility to improve CO.
- **Vasodilators:** Reduce afterload to decrease the heart's workload.
- **Neurohormonal Blockers:** Prevent complications from chronic sympathetic stimulation and RAAS activation (e.g., beta-blockers, ACE inhibitors).
- **Antiarrhythmics:** Treat or prevent ventricular/atrial tachyarrhythmias to maintain cardiac hemodynamics and prevent sudden death.
- **Anticoagulants/Thrombolytics:** Dissolve or prevent thrombi secondary to heart disease.
- **Other Medications:** Low-sodium diet, Omega-3 fatty acids, taurine/carnitine supplementation, cough suppressants.

2.1 Medications Used in Heart Failure

1. Diuretics

Diuretics are the cornerstone of therapy for congestion and fluid retention in heart failure. They reduce circulating blood volume, decrease venous return, and consequently reduce ventricular filling pressures (preload). This reduction in pressure lowers hydrostatic pressure in pulmonary and systemic veins, thereby alleviating and preventing pulmonary edema and ascites.

However, the use of diuretics activates the renin-angiotensin-aldosterone system (RAAS), hence monotherapy is not recommended. Careful titration is essential. While appropriate diuretic use prevents edema without significantly reducing stroke volume (SV), excessive administration can drastically decrease SV, worsening heart failure and leading to prerenal azotemia. Therefore, the goal is always to use the **minimum effective dose** to prevent edema while monitoring for adverse effects. Conversely, withholding diuretics from patients who need them can increase cardiac burden and impair vital organ function.
Commonly used diuretics in small animal practice include:

- **Loop Diuretics:** Furosemide, Torsemide (most potent).
- **Thiazide Diuretics:** Hydrochlorothiazide.
- **Potassium-Sparing Diuretics (Aldosterone Antagonists):** Spironolactone.

Diuretics are used to:

1. **Decrease preload** (alleviate congestion by reducing cardiac overload).
2. **Lower blood pressure** (though rarely used for this purpose in animals).
3. **Treat congestive heart failure** (alleviate symptoms like acute pulmonary edema and systemic congestion).

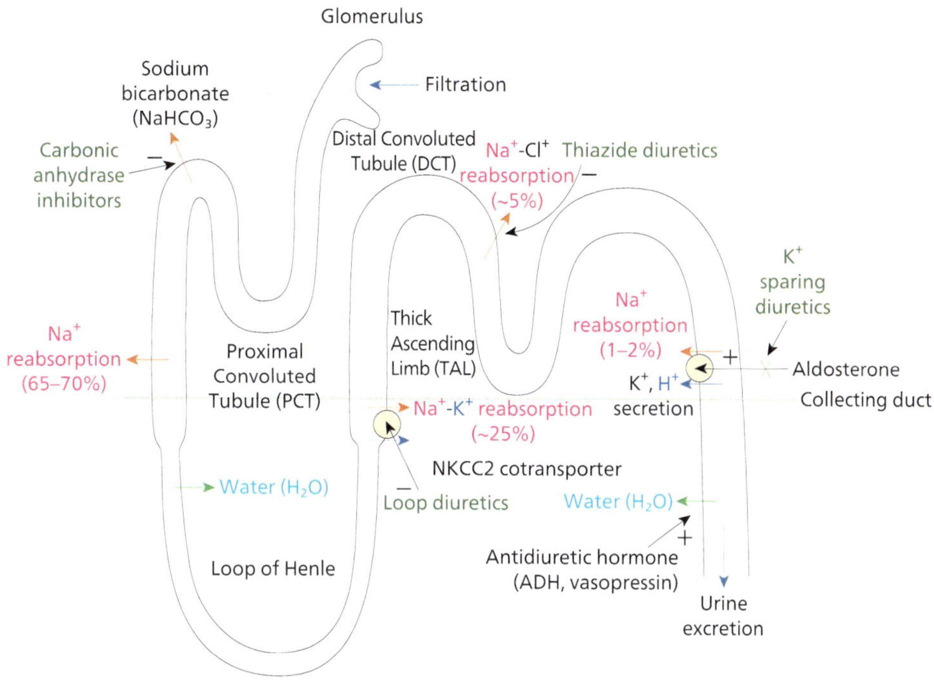

Figure 2-1 ▸ Site of Action for Each Diuretic. Loop diuretics block the sodium pump in the thick ascending limb of the loop of Henle. Thiazide diuretics block the sodium-chloride transporter in the distal convoluted tubule. Potassium-sparing diuretics inhibit aldosterone function in the collecting tubule, suppressing sodium-potassium exchange, thereby exerting diuretic and potassium-sparing effects. PCT (proximal convoluted tubule), DCT (distal convoluted tubule), TAL (thick ascending limb).

Crucially, diuretics are not indicated for asymptomatic cardiac disease patients, as they do not prevent disease progression and carry the risk of renal side effects.

Mechanism of Action: Diuretics promote diuresis by inhibiting sodium reabsorption in different parts of the renal tubules. Combining different classes of diuretics can achieve synergistic effects.

Loop Diuretics (Furosemide, Torsemide):

- Act on the thick ascending limb of the loop of Henle,the most crucial site for sodium reabsorption, by inhibiting the Na-K-2Cl cotransporter (NKCC2), leading to powerful diuretic effects.
- Also promote renal prostaglandin synthesis, increasing renal blood flow and venous capacitance.
- Generally safe, but can cause **hypokalemia, hypocalcemia, and metabolic alkalosis**.
- Concurrent use with steroids can worsen hypokalemia, and hypokalemia can exacerbate digitalis toxicity.

- May increase ototoxicity and nephrotoxicity when combined with aminoglycosides, and nephrotoxicity when combined with NSAIDs.
- Can activate the RAAS and sympathetic nervous system; therefore, co-administration with ACE inhibitors is recommended.
- Cats are more susceptible to loop diuretic side effects, requiring careful dose adjustment and renal function monitoring.
- Combining low-sodium diets with diuretics can paradoxically activate RAAS; consider co-administering ACE inhibitors unless severe heart failure is present.

Furosemide:

- First-line and widely used for fluid retention (e.g., pulmonary edema) in heart failure patients. It also acutely increases venous capacitance.
- Administered orally, via bolus injection, or continuous rate infusion (CRI).
- The dose must be titrated to the minimum effective amount to control clinical signs. Excessive use risks prerenal azotemia and electrolyte disturbances.Because it activates the RAAS, monotherapy is not advised.
- Suffers from a **ceiling effect** and has lower oral bioavailability than torsemide.

Torsemide:

- Torsemide is approximately 10 times more potent than furosemide. It shows decreased susceptibility to diuretic resistance and sustained efficacy with long-term administration.
- It is more bioavailable and has a longer duration of action than furosemide, with adjunctive anti-aldosterone properties.
- It is a good alternative to furosemide, often reserved for refractory congestive heart failure (CHF), potentially avoiding dose escalation.
- Initial oral dose for furosemide: 1-2 mg/kg q12h; for torsemide: 0.1-0.2 mg/kg q12-24h.
- Doses can be gradually increased (furosemide up to 4 mg/kg q8h, max 12 mg/kg/day; torsemide up to 0.4 mg/kg q12h, max 1.2 mg/kg/day).
- If congestion persists at maximal doses of furosemide or torsemide, consider adding a second diuretic like spironolactone or hydrochlorothiazide (**sequential nephron blockade**).

Thiazide Diuretics (Hydrochlorothiazide):

- Second most common diuretics, acting on the distal convoluted tubule to block sodium-chloride transporters.

- Less potent than loop diuretics (about 50% efficacy) because less sodium reabsorption occurs at this site.
- Primarily used as adjuncts to loop diuretics in refractory congestive heart failure (sequential nephron blockade) or to increase renal calcium reabsorption to inhibit calcium stone formation.
- Can cause **hypokalemia** and **metabolic alkalosis** by increasing sodium delivery to the distal tubule, promoting hydrogen and potassium excretion.
- Also activate RAAS, necessitating co-administration of ACE inhibitors.
- Concurrent use with beta-blockers can worsen dyslipidemia and hyperglycemia.
- Recommended doses: Dogs (1-5 mg/kg IM/PO q12h with low-sodium diet), Cats (12.5 mg/head IM, 2-4 mg/kg PO q12h with low-sodium diet).

Potassium-Sparing Diuretics (Aldosterone Antagonists - Spironolactone):

- Unlike other diuretics, they don't directly act on sodium transport but antagonize aldosterone's action in the collecting duct.
- Inhibit sodium (and water) reabsorption without affecting potassium or hydrogen ion excretion, hence "potassium-sparing."
- Have very weak diuretic effects (about 10% of loop diuretics) due to the low sodium reabsorption at this site.
- Rarely used alone; often combined with thiazide or loop diuretics to counteract hypokalemia.
- Can cause **hyperkalemia**, especially when co-administered with ACE inhibitors.
- Recent research suggests antioxidant, anti-fibrotic, and anti-proliferative extrarenal effects, potentially slowing HF progression.
- Long-term use can cause ulcerative facial dermatitis, primarily in certain Maine Coon cats, which resolves upon discontinuation.
- Spironolactone dose for dogs and cats: 1-2 mg/kg PO q12h or 2-4 mg/kg PO q24h.
- Studies on its effect in preclinical MMVD have shown limited clear improvement, but spironolactone groups exhibited less increase in echocardiographic heart size and cardiac biomarker concentrations compared to placebo.

Long-term Diuretic Use Side Effects:

- Volume depletion or reduced renal perfusion can lead to prerenal azotemia.
- **Electrolyte imbalances** are common, including hypokalemia, hyponatremia, and hypochloremia (metabolic alkalosis).
- Continuous monitoring is vital to use the minimum dose.
- Intravenous fluids should generally be avoided in volume-overloaded heart failure

patients, as this increases the risk of pulmonary edema. If severe azotemia persists, judicious fluid therapy (e.g., half-strength saline) may be initiated.

- Hypokalemia can increase the risk of cardiac arrhythmias and digitalis toxicity. Hypokalemia from loop diuretics often normalizes after 10-14 days due to compensatory sodium retention. ACE inhibitors can mitigate hypokalemia.
- If azotemia worsens significantly due to diuretics, consider reducing diuretic dose and increasing vasodilators or inotropes.
- For persistent hypokalemia, oral potassium supplementation can help.
- Regular monitoring of BUN, creatinine, Na, K, and Cl is crucial to adjust doses.

2. RAAS Blockers (Renin-Angiotensin-Aldosterone System Blockers)

Angiotensin-Converting Enzyme Inhibitors (ACEi):

- Inhibit the formation of angiotensin II, a potent vasoconstrictor produced by renin, leading to vasodilation.
- Increase bradykinin levels (a vasodilator) by inhibiting ACE, which normally breaks down bradykinin. This is a suspected cause of dry cough in humans but is rarely observed in dogs and cats.
- Primary actions: Balanced vasodilation (acting on both arteries and veins), reducing systemic arterial and venous pressure, thereby lowering ventricular preload and afterload.
- Promote renal sodium and water excretion, reducing circulating volume, and inhibit sympathetic nervous system activity. ACE inhibition is generally incomplete due

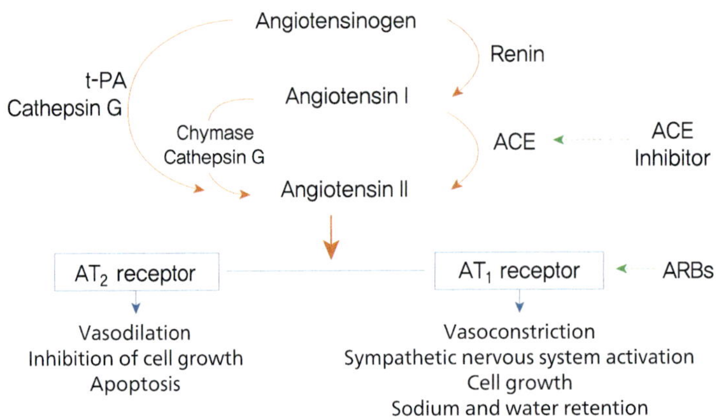

Figure 2-2 ▸ Effects of Angiotensin II on the Circulatory System Angiotensin II (AII), formed by the Renin-Angiotensin-Aldosterone System (RAAS), induces cardiac and vascular hypertrophy, systemic vasoconstriction, increased circulating blood volume, and increased renal sodium and water retention. ACE inhibitors (ACEi) block these effects of AII.

to alternate synthesis pathways (e.g., chymase).

- In humans, long-term ACEi use is proven to inhibit myocardial and vascular hypertrophy, delaying cardiac remodeling. This effect is not yet definitively proven in small animals.
- Clinically used for HF, hypertension secondary to renal disease, and proteinuria.
- Side effects: Rare angioedema, azotemia, and hyperkalemia.
- Can induce acute kidney injury in dehydrated patients or those receiving nephrotoxic drugs (NSAIDs, aminoglycosides).
- Contraindicated in pregnant and lactating animals.

ACEi Efficacy in Dogs with Heart Failure:

- Large-scale studies (SVEP 2002, VETPROOF 2002) in asymptomatic dogs revealed limited effect of ACEi in preventing or slowing heart failure progression due to MMVD.
- Although the final VETPROOF study (2007) data showed a tendency toward a favorable effect in ISACHC Ib (ACVIM B2, asymptomatic with radiographic cardiomegaly) patients, definitive evidence that ACEi improve prognosis in subclinical MMVD is lacking.
- The 2019 ACVIM guidelines for canine MMVD allow ACEi use in asymptomatic patients with cardiomegaly.
- Studies in symptomatic DCM and MMVD patients (COVE 1995, Live 1998, BENCH 1999) demonstrated that ACEi slowed HF progression, improved clinical signs, and prolonged survival.

Common ACEi and Recommended Doses:

- **Enalapril:** Dogs (HF: 0.25-1 mg/kg PO q12-24h; HTN: up to 3 mg/kg); Cats (0.25 mg/kg PO q12-24h).
- **Benazepril:** Dogs (HF: 0.25-0.5 mg/kg PO q24h); Cats (HF: 0.25-0.5 mg/kg PO q24h; Renal failure: 0.5-1.0 mg/kg PO q24h).
- **Imidapril:** Dogs (HF: 0.25 mg/kg PO q24h); Cats (0.25 mg/kg PO q24h - unverified dose).
- **Ramipril:** Dogs (HF: 0.125 mg/kg PO q24h); Cats (0.5 mg/head PO q24h).

Angiotensin Receptor Blockers (ARBs):

- Block type 1 angiotensin II (AT1) receptors in the heart and blood vessels.
- Do not inhibit ACE, thus do not increase bradykinin levels, avoiding side effects like dry cough.

- Pharmacological actions are similar to ACEi: balanced vasodilation (reducing arterial pressure, preload, and afterload), promoting renal sodium and water excretion, and alleviating cardiac congestion.
- In humans, used for hypertension and HF.
- In small animals, efficacy is less established due to limited research, and their use is restricted as ACEi side effects are less prevalent than in humans.
- Telmisartan and Valsartan are examples being utilized clinically for proteinuria and systemic hypertension.
- Use with caution in patients at risk of renal failure or receiving nephrotoxic drugs, as ARBs can precipitate azotemia by lowering glomerular filtration pressure.

Aldosterone Breakthrough (ABT):

- Refers to the ineffective blockade of the RAAS despite ACEi or ARB therapy, a major concern with long-term ACEi use.
- This effect is exacerbated by loop diuretics, which further activate the RAAS, and/or severe dietary sodium restriction, which stimulates aldosterone synthesis and secretion.
- Reported in up to 60% of humans on long-term ACEi/ARB and about 30% of dogs on cardiac medications.
- To prevent ABT, avoid overly rapid low-sodium diets and use minimal effective loop diuretic doses.
- If ABT occurs and cannot be controlled by the above methods, consider increasing ACEi dose or adding spironolactone.

3. Vasodilators

Vasodilators relax vascular smooth muscle, dilating blood vessels.

- **Arterial vasodilators** reduce systemic vascular resistance (afterload) and arterial blood pressure, lessening ventricular pressure overload.
- **Venous vasodilators** increase vascular capacitance, reducing venous return (VR) and consequently ventricular preload, alleviating volume overload.
- Used primarily for HF and hypertension. Most are mixed vasodilators, but some are pure arterial or venous.
- Some calcium channel blockers (CCBs) are also used for tachyarrhythmias or to improve myocardial relaxation.
- Examples include Hydralazine (pure arterial), Nitrates (venous or mixed), Pimobendan (mixed inodilator), alpha-1 blockers(mixed), and ACE inhibitors (mixed).

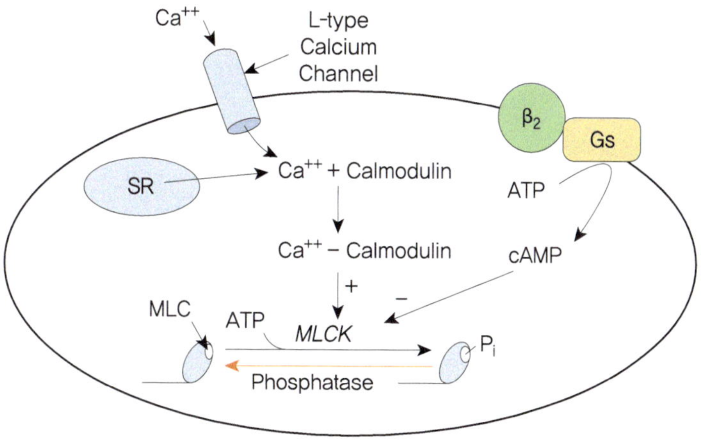

Abbreviations; SR. sarcoplasmic reticulum; gq, Gs-protein; MLC, myosin light chain; MLCK, mysin light chain kinase; P_i, myosin phosphorytation

Figure 2-3 ▸ Mechanism of Action of Calcium Channel Blockers. CCBs block L-type calcium channels, leading to negative inotropic effects on the heart and relaxation of vascular smooth muscle, resulting in vasodilation. L-type calcium channels are distributed in vascular smooth muscle, cardiomyocytes, and the SA and AV nodes. Blocking them slows the heart rate (used as a treatment for tachyarrhythmias), increases myocardial relaxation, and dilates blood vessels.

Alpha-Adrenoceptor Antagonists (e.g., Prazosin):

- Block alpha-sympathetic receptors on vascular smooth muscle, causing vasodilation.
- Dilate both arteries and veins, with a stronger effect on arteries.
- Effective in conditions with increased sympathetic tone.
- Selective alpha-blockers like prazosin are used for primary hypertension in humans but are not widely used in small animals.
- Side effects: Hypotension and reflex tachycardia.
- Not shown to be effective for HF in humans, so not used for this purpose.
- Concurrent use with beta-blockers or diuretics can cause severe hypotension.

Calcium-Channel Blockers (CCBs):

- Block L-type calcium channels, leading to negative inotropic effects on the heart and relaxation of vascular smooth muscle.
- L-type calcium channels are found in vascular smooth muscle, myocardial cells, SA node, and AV node.
- Reduce heart rate (used for tachyarrhythmias) and slow atrioventricular conduction, improve myocardial relaxation (lusitropy), and cause vasodilation.
- Pharmacological effects vary by CCB type.
- In small animals, primarily used for diseases causing myocardial hypertrophy, atrial

arrhythmias, and hypertension.

- **Dihydropyridine CCBs (Amlodipine, Nifedipine):** High affinity for blood vessels; primarily used for systemic hypertension. They can cause excessive hypotension and reflex tachycardia. Amlodipine may be used adjunctively for refractory heart failure.
- **Non-dihydropyridine CCBs (Diltiazem, Verapamil):** Act preferentially on nodal tissue, resulting in decreased sinus rate and reduced atrioventricular conduction velocity. They are beneficial for myocardial hypertrophy (e.g., improving diastolic function in cats with HCM) and alleviate atrial arrhythmias.
- Rarely but when indicated, CCBs may be **co-administered with beta-blockers (BBs)** for enhanced antiarrhythmic effects, but this requires caution due to the risk of inducing severe bradycardia, heart block, and hypotension.

Hydralazine:

- Hydralazine is a direct systemic arterial vasodilator, which acts upon vascular smooth muscle and reduces peripheral resistance and blood pressure (BP).
- Hydralazine is a semicarbazide-sensitive amine oxidase (SSAO) inhibitor. It is believed that hydralazine alters cellular calcium metabolism in smooth muscle, thereby interfering with calcium movements and preventing the initiation and maintenance of the contractile state.
- Promotes nitric oxide formation in vascular endothelial cells, leading to cGMP-mediated vasodilation.
- Primarily used for arterial hypertension.
- Short half-life, requiring frequent administration, and can cause reflex tachycardia, so not used for chronic hypertension (acute hypertension only).
- Can be used for HF due to valvular insufficiency by reducing regurgitant blood flow, but reserved for refractory cases due to hypotension risk.
- Major side effects: Reflex tachycardia, hypotension, and gastrointestinal disturbances. The use of hydralazine in cats has generally fallen out of clinical favor.Lupus-like symptoms seen in humans are not reported in small animals.

Nitrate Vasodilators:

- Release nitric oxide (NO) to primarily dilate systemic veins.
- NO, produced by endothelial cells, relaxes vascular smooth muscle (vasodilation), inhibits platelet aggregation (antiplatelet effect), and interferes with leukocyte-endothelial cell interaction (anti-inflammatory).
- These effects occur via cGMP increase.
- Drugs releasing NO spontaneously(e.g., sodium nitroprusside) or organic nitrates (e.g., nitroglycerin, isosorbide dinitrate) are used.

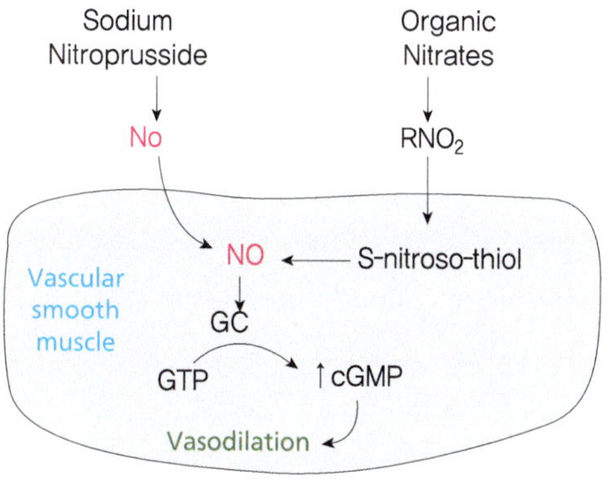

Figure 2-4 ▸ Mechanism of Action of Nitrate Vasodilators. Nitric oxide (NO) is primarily produced by vascular endothelial cells and has the effect of relaxing vascular smooth muscle, leading to vasodilation. This effect occurs as NO breaks down GTP, increasing cGMP levels, which in turn relaxes vascular smooth muscle and causes vasodilation. Sodium nitroprusside is a drug that spontaneously releases NO, while organic nitrates release NO through enzymatic breakdown.

- Tolerance can develop with continuous administration (e.g., more than 36 to 72 hours).
- Are **balanced vasodilators** (affect both arteries and veins), but venous dilation predominates at normal doses.
- Systemic venous dilation reduces ventricular preload, alleviating volume overload and edema, and decreases ventricular wall stress and oxygen demand, reducing cardiac workload.
- High doses can cause systemic vasodilation, hypotension, and reflex tachycardia.
- Used for hypertension and HF. Administration route, onset, and duration vary by specific nitrate.

Nitroglycerin:

- Rapid onset (2-5 minutes), short duration (30 minutes), so used most often for the short-term management of acute CHF.
- Transdermal patches used for longer effect; not used for >12 hours to prevent tolerance (e.g., 12h On/12h Off).
- Efficacy in dogs and cats when applied topically is inconsistent due to variable absorption through hair-covered skin. Sublingual administration is not practical.

Isosorbide Dinitrate (ISDN):

- Tolerance can develop with continuous use (e.g., more than 36 hours); this may be avoided by scheduling a drug-free period (e.g., 12 or 24 hours on and 24 hours off).
- It can be combined with hydralazine for balanced vasodilation.

- Longer onset and duration than nitroglycerin.
- Can be given orally, but higher doses are needed compared to sublingual or transdermal.
- Considered justifiable for chronic recurrent pulmonary edema given its role in acute pulmonary edema treatment (furosemide, nitrate, oxygen, dobutamine).
- Often overlooked in veterinary medicine due to short half-life and oral administration difficulties.
- Clinical experience suggests ISDN is effective for recurrent pulmonary edema, but strict blood pressure monitoring is vital to avoid severe azotemia from reduced renal perfusion.

Sodium Nitroprusside:

- Potent, direct-acting mixed vasodilator (arterial and venous). Used for short-term treatment of refractory, life-threatening congestive heart failure or critical systemic hypertension.
- Rapid onset, short half-life,administered as an IV continuous rate infusion and requires arterial blood pressure monitoring due to the risk of profound hypotension.
- Major adverse effects include hypotension, tachycardia, nausea, and vomiting.
- Long-term use carries risk of thiocyanate toxicity.
- Not practical for primary care clinics. Requires light protection during administration.
- Patients rapidly develop tolerance, limiting long-term use.

Recommended Vasodilator Doses (examples):

- **Prazosin:** Dogs (1.0 mg PO q8-12h for <15kg, 2.0 mg PO q8-12h for >15kg); Cats (0.25-1.0 mg PO q8-12h).
- **Amlodipine:** Dogs (0.05-0.1 mg/kg PO q12-24h, titrate weekly based on BP); Cats (0.625-1.25 mg/head PO q24h, titrate as needed, BID possible).
- **Verapamil:** Dogs (0.5-3 mg/kg PO q8h; 0.05 mg/kg IV over 10 min, up to 4 doses, then 50% reduction); Cats (0.5-1 mg/kg PO q8h; 0.025 mg/kg IV over 5 min, up to 8 doses).
- **Diltiazem:** Dogs (0.5-2.0 mg/kg PO q8h); Cats (0.5-2.5 mg/kg PO q8h, or 10 mg tablet/head PO q8h).
- **Hydralazine:** Dogs (initial 0.5-3 mg/kg PO q8-12h, titrate to 2-3 mg/kg q12h with BP monitoring); Cats (2.5-10 mg/head PO q12h, titrate carefully).
- **Nitroglycerin:** Dogs (1/4 patch/5kg on ear pinna, axilla, abdomen, groin; 12h On/12h Off due to tolerance; sustained-release 2.6-6.4 mg PO q8-12h); Cats (1/4 patch/head, sustained-release 2.6 mg PO q8-12h).

- **Sodium Nitroprusside:** Dogs/Cats (1-15 μg/kg/min IV CRI, start at 1 μg/kg/min, increase 0.5-1 μg/kg/min every 5 min until effect, most controlled by 3 μg/kg/min; discontinue if systolic BP $<$100 mmHg).
- **Isosorbide Dinitrate:** Dogs (0.5-2 mg/kg PO q8-12h); Cats (2.5 mg/head PO q12h).

4. Inotropes

Positive inotropic agents increase myocardial contractility, thereby increasing cardiac output and arterial pressure. These drugs have variable effects on heart rate (chronotropy). Some agents increase heart rate (e.g., sympathomimetics), while others, like Digoxin, slow heart rate by vagal effects. Some agents cause systemic vasodilation (e.g., Pimobendan), while others induce vasoconstriction (e.g., high doses of Dopamine). They are used for heart failure (HF), cardiogenic shock, and hypotension.

- **Acute HF (hospitalized patients with sudden worsening):** Dobutamine is primarily used.
- **Chronic HF (managed at home):** Pimobendan is used.
- **Other drugs in dogs:** Digoxin (often for supraventricular arrhythmias due to parasympathetic effects) and Milrinone (phosphodiesterase inhibitor, less common).

Digitalis (Cardiac Glycosides - Digoxin):

- Isolated from foxglove (Digitalis purpurea), used since 1785.
- In dogs, Digoxin is a relatively weak positive inotrope, and as such, it is not useful in the emergency setting. It is mainly used for supraventricular tachyarrhythmias or atrial fibrillation , particularly for ventricular rate control.
- In cats, Digoxin has a very narrow therapeutic range, and intravenous administration is not recommended due to the high likelihood of toxicity. Toxicity is potentiated by hypokalemia, hyponatremia, hypercalcemia, hypothyroidism, and hypoxia.
- **Mechanism:** Inhibits myocardial sarcolemmal Na^+/K^+-ATPase, indirectly inhibiting Na^+-Ca^{2+}-exchanger, increasing intracellular calcium concentration. This increases calcium availability for troponin-C, enhancing myocardial contractility.
- Acts as a negative chronotrope by central vagomimetic effects, which slow heart rate and AV node conduction. This effect is useful for slowing the ventricular response in atrial tachyarrhythmias, such as AF.
- Values reported for the elimination half-life of digoxin in dogs and cats have been highly variable, with values reported from 30 to 42 hours for dogs. It requires 3 to 5 days to reach an effective steady-state concentration with maintenance dosing. Rapid IV or oral digitalization is generally not recommended due to the high likelihood of toxicity.
- The therapeutic trough plasma concentration is typically 0.5-1.5 ng/mL, measured 8

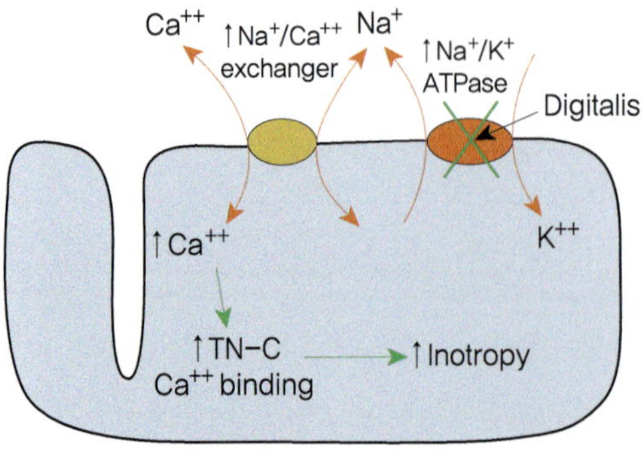

Figure 2-5 ▶ Mechanism of Action of Digitalis. Digitalis blocks the Na^+/K^+-ATPase on the myocardial sarcolemma, inhibiting calcium release from the Na^+-Ca^{2+}-exchanger. This increases intracellular calcium concentration, leading to more calcium binding with troponin-C, thereby enhancing myocardial contractility.

to 12 hours after dosing.

- Toxicity (digoxin levels >2.0 ng/mL): Ventricular arrhythmias, AV block.
- Immune Fab (Digibind) can be used to lower digoxin levels.
- Digoxin toxicity initially manifests as gastrointestinal signs (inappetence, vomiting, nausea, diarrhea), which almost always precede cardiac arrhythmias when the drug is administered orally. Toxicity is exacerbated by hypokalemia, hypothyroid states, and hypercalcemia. Toxic cardiac signs include ventricular arrhythmias (e.g., ventricular premature complexes, ventricular tachycardias) and various degrees of AV block.
- Wide individual variability in dogs, making therapeutic range difficult to maintain. Cats are more sensitive and prone to toxicity.
- Class IA antiarrhythmics compete with digoxin for binding sites, decreasing renal digoxin clearance and increasing toxicity risk. CCBs and NSAIDs also increase toxicity risk.
- Renal dysfunction reduces digoxin excretion, increasing toxicity risk.
- Commonly used: Digoxin, Digitoxin.
- Recommended Digoxin doses: Dogs (2.2-4.4 μg/kg IV q12h, 5.5-11 μg/kg PO q12h; >22 kg dogs: 0.22 mg/m2); Cats (1-1.6 μg/kg IV q12h, 10 μg/kg PO q24-48h, 1/4 of 125 μg tablet q24-48h).

Beta-Agonists (Dobutamine, Dopamine):

- Stimulate β1-receptors in myocardium and smooth muscle, increasing heart rate and myocardial contractility.
- Used for HF and circulatory shock.
- Almost all beta-agonists have β1-agonist effects. Dose-dependent effects.

- Cardiac effects include positive inotropy and tachycardia, though Dobutamine is often preferred for its ability to increase contractility without significantly increasing heart rate.
- Vascular effects are dose-dependent: beta 1 and beta 2 stimulation (Dobutamine) often results in no major change in blood pressure, while Dopamine may cause deleterious vasoconstriction due to alpha-adrenergic effects at higher doses.
- Can cause dangerous ventricular tachyarrhythmias and increase myocardial oxygen demand with long-term use, potentially accelerating myocardial remodeling.
- Due to side effects, primarily used for acute HF (Dopamine, Dobutamine), administered via CRI due to short half-life.
- Recommended Doses (CRI, syringe pump required):
    - **Dobutamine:** Dogs (5-20 μg/kg/min CRI); Cats (2-5 μg/kg/min CRI). Discontinue if heart rate becomes too high (>200 bpm in dogs, >250 bpm in cats) or arrhythmias occur. ECG monitoring required. Max 24-48 hours.
    - **Dopamine:** Dogs (2-10 μg/kg/min CRI); Cats (1-5 μg/kg/min CRI). Discontinue if heart rate too high or arrhythmias occur. ECG monitoring required.

Phosphodiesterase Inhibitors (PDEi):

- Inhibit phosphodiesterase, the enzyme that breaks down cAMP, increasing intracellular cAMP levels.
- Increased cAMP in myocardial cells enhances contractility, increases heart rate, and accelerates conduction velocity.
- Used in HF for their inotropic and unique vasodilatory effects.
- **cAMP-dependent PDE3 inhibitors (PDEi-3):** Milrinone, Amrinone, Pimobendan, Levosimendan. These are inodilators (increase contractility and reduce systemic vascular resistance).
- **cGMP-dependent PDE5 inhibitors (PDEi-5):** These agents act by accumulation of cGMP in vascular smooth muscle, resulting in vasodilation. They primarily dilate the pulmonary vasculature and are used for pulmonary hypertension (and erectile dysfunction in human).
- In dogs, Pimobendan has demonstrated increased survival times and improved quality of life in patients with CHF secondary to DCM or chronic valvular disease. The bipyridines, Milrinone and Amrinone, are limited to short-term IV use for severe refractory myocardial failure.

Pimobendan:

- **Inodilator** with positive inotropic and vasodilatory effects.

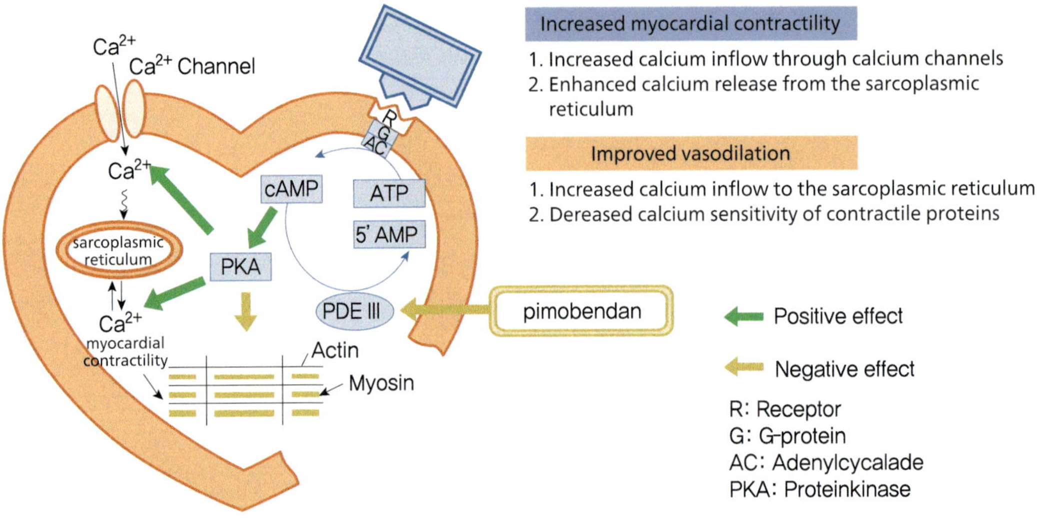

Figure 2-6 ▸ Pharmacological Effects of Pimobendan. Pimobendan is an inodilator that possesses both positive inotropic and vasodilatory effects. In addition to its PDEi-3 effect, it also acts as a calcium sensitizer by increasing the sensitivity of intracellular contractile elements to calcium, thereby exhibiting a cardiotonic effect. Pimobendan inhibits vasoconstrictors like PDE III, relaxing vascular smooth muscle, reducing peripheral vascular resistance, and decreasing afterload.

- **Mechanism:** Beyond PDEi-3 effect, it increases calcium sensitivity of myocardial contractile elements, acting as a "calcium sensitizer."
- Increases contractility without increasing myocardial oxygen demand, avoiding ischemic changes in the failing heart.
- Increases contractility with minimal arrhythmia risk, unlike digitalis.
- Inhibits vasoconstrictors like PDE3, relaxing vascular smooth muscle, reducing peripheral vascular resistance, and decreasing afterload.
- Weakens expression of pro-inflammatory mediators and iNOS, delaying myocardial inflammatory damage and improving systolic dysfunction.
- Reduces norepinephrine levels (which are elevated in HF), thereby inhibiting norepinephrine-mediated HF progression.
- Increases heart rate in normal dogs with volume overload but has little effect on heart rate in HF animals, thus no arrhythmia induction like digitalis.
- Available as chewable tablets and capsule(Vetmedin®).
- Can be co-administered with diuretics, ACEi, and digoxin.
- Positive inotropic effect may be reduced when co-administered with CCBs or beta-blockers.
- Capsules should be given at least 1 hour before feeding due to gastric acid/food effects, but chewable tablets can be given regardless of food. Primarily oral. IV use for emergencies in Europe, but strong inotropic effects like dobutamine are not seen.

Pimobendan Indications:

- Chronic treatment of all patients with HF secondary to MMVD or DCM.
- Can be used for other heart diseases unless contraindicated (e.g., cardiac tamponade, left ventricular outflow tract obstruction).
- Proven to have significant beneficial effects in symptomatic HF patients, improving clinical signs and prolonging survival (PITCH 2000, QUEST 2008). Effects are particularly pronounced in MMVD and DCM.
- Some clinical improvement observed in ascites or pulmonary hypertension from heartworm disease.
- Longitudinal Analysis(2018) of EPIC study(2016) investigated pimobendan in preclinical (asymptomatic) MMVD patients. It showed a reduction in LVIDd and LA/Ao at day 35 in the pimobendan group compared to placebo, although CHF onset was similar between groups.
- ACVIM guidelines recommend pimobendan for ACVIM B2 (ISACHC Ib) patients (asymptomatic with cardiomegaly) based on this study.
- The author recommends early pimobendan therapy for all patients with cardiomegaly, always co-administered with ACEi.

Recommended PDEi Doses:

- **Pimobendan:** Dogs (0.25-0.5 mg/kg PO q8-12h, 1 hr before food for capsules, chewable regardless of food); Cats (1.25 mg/head PO q12h, 1 hr before food for capsules, chewable regardless of food).
- **Milrinone:** Dogs (1-10 μg/kg/min CRI).
- **Sildenafil:** Dogs (0.5-3 mg/kg PO q6-8h).

5. Beta-Blockers

Beta-blockers inhibit sympathetic nervous system activity, reducing heart rate and myocardial contractility, thereby decreasing cardiac output and arterial pressure. This alleviates myocardial workload and reduces myocardial oxygen consumption. They also slow cardiac conduction and improve myocardial relaxation.

- Primarily used for **hypertrophic cardiomyopathy (HCM)** or other diastolic myocardial diseases causing myocardial hypertrophy, and for atrial arrhythmias with abnormally fast AV nodal conduction.
- The heart has β1 and β2 receptors, with β1 being functionally and numerically dominant. These receptors bind norepinephrine, causing palpitations. Beta-blockers competitively bind to these receptors, blocking norepinephrine's effects.
- In human heart disease, chronic sympathetic activation contributes to cardiac re-

modeling and worsens HF. Beta-blockers are proven to improve clinical signs, slow HF progression, and prolong survival in humans.

- **Unfortunately, these effects are not yet proven in canine and feline heart disease.**
- **Non-selective beta-blockers** (Propranolol) block both β1 and β2 receptors, used for cardiomyopathy.
- **Selective β1-blockers** (Atenolol, Esmolol) block only β1, primarily used for arrhythmias.
- Newer, broader-spectrum beta-blockers (Carvedilol) also block alpha-receptors. However, Carvedilol's efficacy in HF has not been clinically proven.
- ACVIM guidelines recommend cautious use of beta-blockers and advise against their use in patients with pulmonary edema.
- Beta-blockers require careful **upward titration dosing** and should not be started or stopped abruptly.
- Studies in canine DCM and MMVD have not shown improved clinical signs or survival rates with beta-blockers. No beneficial effects observed in asymptomatic heart disease patients. Some studies even suggest avoiding their use in dogs with overt HF symptoms.
- **In both dogs and cats, beta-blockers are indicated when tachyarrhythmias contribute to HF symptoms.**

Recommended Beta-Blocker Doses:

- **Propranolol:** Dogs (0.02-0.08 mg/kg IV slow over 5 min q8h, 0.25-1.5 mg/kg PO q8h); Cats (0.04-0.06 mg/kg IV slow over 5 min q8h, 2.5-5 mg/head PO q8h).
- **Atenolol:** Dogs (0.25-1.5 mg/kg PO q12h); Cats (5-12.5 mg/head PO q24h, then q12h after 1 week).
- **Esmolol:** Dogs/Cats (0.05-0.1 mg/kg IV 5 min apart until cumulative 0.5 mg/kg; 50-200 μg/kg/min CRI).
- **Carvedilol:** Dogs (start 0.1-0.5 mg/kg PO q12h, titrate every 2 weeks up to 1.5 mg/kg).

6. Antiarrhythmics

Arrhythmias can both cause and be caused by heart disease. Antiarrhythmics are chosen based on whether the arrhythmia is bradycardic or tachycardic.

- **Bradyarrhythmias:** Treated with sympathetic stimulants or vagolytic agents depending on the cause.
- **Tachyarrhythmias:** Treatment depends on the origin (type).
  - **Supraventricular arrhythmias** (atrial, junctional - originating above the AV junction): Diltiazem (Class IV, calcium channel blocker) or Atenolol (Class II, beta-blocker) are first-line.

- **Ventricular tachyarrhythmias** (originating below the AV junction): Lidocaine (Class I, sodium channel blocker) for emergencies, and Sotalol or Amiodarone (Class III, potassium channel blockers) for chronic management.

Important Considerations:

- Antiarrhythmics have **negative inotropic effects** and can cause debilitation; choosing the right time to initiate treatment is clinically important.
- Infrequent ventricular premature contractions (VPCs) that do not cause clinical signs or sudden death are often observed without treatment.
- Most atrial arrhythmias without causing sustained tachycardia are benign and treatment focuses on the underlying heart disease, but conditions like atrial fibrillation causing restlessness or syncope require immediate treatment.

Recommended Antiarrhythmic Doses (examples):
[Atrial Tachyarrhythmia Treatment]

- **Digoxin:** Dogs (0.22 mg/m2 PO q12h); Cats (0.125 mg tablet 1/4-1/8 PO q24-48h).
- **Propranolol (Class II):** Dogs (0.21-1.0 mg/kg q8h PO; 0.04-0.06 mg/kg IV slow); Cats (0.21-1.0 mg/kg q8-12h PO; 0.04 mg/kg IV slow).
- **Metoprolol (Class II):** Dogs (0.5-1 mg/kg q8h PO); Cats (2-15 mg/kg q8h PO).
- **Atenolol (Class II):** Dogs (0.5-2 mg/kg q12h PO); Cats (5-12.5 mg/head q24h PO).
- **Diltiazem (Class IV):** Dogs (0.5 mg/kg TID PO); Cats (0.5 mg/kg PO q8h for atrial fibrillation, 7.5 mg/head PO q8-12h for HCM).

[Ventricular Tachyarrhythmia Treatment]

- **Lidocaine (Class IB, emergency ventricular tachyarrhythmia):** Dogs (2-4 mg/kg IV slow (2% lidocaine diluted 10x), repeat every 10 min up to 8 mg/kg; CRI: 25-75 μg/kg/min); Cats (0.25-1.0 mg/kg IV slow over 5 min).
- **Tocainide (Class IB):** Dogs (5-10 mg/kg q6-8h PO).
- **Mexiletine (Class IB):** Dogs (2-5 mg/kg q8-12h PO).
- **Procainamide (Class IA):** Dogs (6-8 mg/kg IV over 5 min; 25-40 μg/kg/min CRI; 8-20 mg/kg q4-6hr IM, q8h PO); Cats (8-20 mg/head q8h PO).
- **Quinidine (Class IA):** Dogs (gluconate 6-20 mg/kg q6h IM; Cardioquin 8-20 mg/kg q6-8h PO); Cats (sulfate 5.5-11.0 mg/kg q8h PO).
- **Sotalol (Class III):** Dogs (0.5-2 mg/kg q12h PO).
- **Amiodarone (Class III):** Dogs (initial: 10-15 mg/kg q12h PO for 1 week, then 5-7.5 mg/kg q12h for 2 weeks, then 7.5 mg/kg q24h; caution with overdose due to cardiac depression and sudden death).

7. Anticoagulants/Thrombolytics

These drugs are categorized as **anticoagulants** (prevent thrombus formation) and **thrombolytics** (dissolve existing thrombi). It's crucial to remember that anticoagulants do not dissolve existing clots. In dogs, anticoagulants are generally not needed unless the condition is related to heartworm disease.

2.2 Nutritional Therapy for Heart Disease

Heart disease is often chronic, leading to cachexia due to long-term medication-induced anorexia and metabolic disturbances from circulatory impairment. Nutritional therapy aims to:

1. **Supplement adequate calories, essential proteins, fats, and vitamins.**
2. **Minimize medication side effects.**
3. **Alleviate congestive heart failure through sodium restriction.**
4. **Minimize myocardial ischemic damage with antioxidants.**
5. **Address nutritional deficiencies contributing to heart disease.**

Historically (1960s-1970s), low-sodium, protein-restricted, and B-vitamin-supplemented diets were recommended for all HF stages. The 1980s saw debates on the timing of sodium restriction and the emergence of "cardiogenic cachexia." The discovery of the link between taurine deficiency and feline dilated cardiomyopathy (DCM) in the late 1980s raised questions about taurine supplementation benefits in dogs, who can synthesize their own.

Key Nutritional Considerations:

- **Adequate Calorie Intake:** Crucial as HF patients have increased cardiovascular energy expenditure but reduced nutrient intake due to stress and poor circulation. Nutritional deficiencies impair immune response and tissue repair, leading to hypoalbuminemia and impaired kidney/liver function, affecting drug metabolism and increasing toxicity risk.
- **Potassium Supplementation:** Long-term diuretic use can cause hypokalemia, often accompanied by severe anorexia, especially when renal disease is co-present. Balanced diets (high in calories and quality protein) and potassium supplementation are vital.
- **Cardiogenic Cachexia (Muscle Mass Loss):** Common in late-stage HF. While obesity was once considered detrimental, human studies show that loss of lean body mass in HF worsens organ function and increases mortality (obesity paradox: slightly overweight patients, BCS 6-7/9, have longer survival). Therefore, **low-fat or low-protein diets are not recommended** for weight loss in HF patients.

- Cardiogenic cachexia affects >50% of HF dogs, involving muscle protein loss, not just fat.
- Supplement with **high-quality protein** (e.g., chicken breast, cottage cheese, egg whites) added to existing food.
- For patients on diuretics, wet food or home-cooked meals are preferred over dry kibble to increase water intake.

- **Improving Appetite in Anorexic Patients:**
  1. Warm food to enhance aroma.
  2. Mix small amounts of human food (avoiding garlic or excess potassium chloride).
  3. Hand-feeding.
  4. Offer small, frequent meals.
  5. Add high-quality fats like fish oil to improve palatability.

- **Omega-3 (N-3) Fatty Acids:** Recommended as inflammatory cytokines (TNF-α, IL-1) mediate cardiogenic cachexia. Fish oil, rich in N-3 fatty acids, can reduce cytokine production, improve cachexia, and enhance appetite in HF dogs. N-3 fatty acid intake can also prolong survival in HF dogs. Alpha-linoleic acid (from plants) is also an N-3 fatty acid.

Omega-3 supplements sold in pet stores may not contain sufficient quantities of the active ingredients. The most commonly used fish oil, available at pharmacies, supermarkets, and health food stores, typically provides approximately **180 mg of EPA and 120 mg of DHA per 1000 mg capsule**. This concentration of fish oil can supply the daily required amount of EPA and DHA for an animal weighing approximately 5 kg. Fish oil supplements should contain **Vitamin E as an antioxidant** but should not include other nutrients to avoid toxicity. **Cod liver oil and flaxseed oil should not be used as primary sources of n-3 fatty acids.**

Sodium Restriction

In animals with heart disease, sodium retention leads to fluid retention, increasing circulating blood volume and raising blood pressure. Therefore, in the 1960s, sodium intake was restricted when signs of heart disease were observed. However, **excessive sodium restriction in early heart disease is not actively recommended** as it can decrease appetite and lead to nutritional deficiencies. Recent studies suggest that sodium restriction can actually **activate the RAAS (Renin-Angiotensin-Aldosterone System) and induce aldosterone breakthrough**, thereby reducing the efficacy of ACE inhibitors (ACEi). Thus, **aggressive sodium restriction is not necessary unless the animal is in a stage of heart failure requiring diuretics.** A study in asymptomatic dogs with MMVD (Canine Myxomatous Mitral Valve Insufficiency) showed that low-sodium diets increased aldosterone levels and heart rate, with no improvement in heart size or function. Consequently, **mild sodium restriction (e.g., <100 mg/100 kcal for ISACHC Ia and Ib patients) is recommended** for asymptomatic patients, as severe restriction has shown no benefits and only adverse effects. This period (before the onset of CHF) is an ideal time to educate owners on dietary management for their pets, as starting a diet is easier and simpler. Sodium restriction should begin when symptoms of heart failure appear, typically requiring **sodium restriction below 80 mg/100 kcal**. At this stage, commercially available prescription cardiac diets or low-sodium regular diets can effectively achieve sodium restriction. Most owners are unaware of the sodium content in pet foods or human foods, so they must be advised to **read ingredient labels before feeding anything**. Encouraging the feeding of vegetables, fruits, and low-sodium pet treats is also recommended. **Prescription diets formulated for kidney disease should not be given to patients with heart disease unless renal disease is also present,** due to their extremely low protein and high-fat content. Sodium restriction can reduce appetite, leading to anorexia.

Effective ways to feed a sodium-restricted prescription diet include:

1. **Gradually introducing the new food:** Start by mixing a few kibbles or rehydrating the original food and mixing it with the rehydrated prescription diet (though rehydrating food is controversial).
2. **Slowly increasing the proportion of the prescription diet over a long period,** typically 2-3 months or more, until the diet is fully transitioned.

Protein

In the 1960s, it was believed that protein intake increased metabolic stress on the kidneys and liver, worsening CHF symptoms, thus protein restriction was recommended. However, the current trend strongly recommends **feeding high-quality protein, including essential nutrients for the heart like taurine and L-carnitine**, rather than excessive protein restriction, due to the adverse effect of chronic disease-induced cachexia and reduced lean body mass.

- **Taurine:** Taurine is an amino acid abundant in cardiac muscle. Taurine deficiency is known to cause **dilated cardiomyopathy (DCM) in cats and some dog breeds** (e.g., Cocker Spaniels). Unlike cats, dogs typically synthesize sufficient taurine, so deficiency is rare. However, low taurine concentrations have been reported to cause DCM in certain breeds (American Cocker Spaniel, Golden Retriever, Labrador Retriever). Also, very low protein diets, lamb/rice-based diets, and some high-fiber diets have been linked to taurine deficiency. Methionine is a precursor to taurine. Liver utilizes methionine and cysteine to form taurine. Diet low in methionine appear to cause taurine deficiency. Taurine deficiency can also result from increased renal or fecal loss, increased requirements, or metabolic disorders identified in some breeds. **Taurine supplementation (250-1000 mg, PO, q8-12h) is necessary for dogs with taurine deficiency** (the exact dose needed for supplementation is still unknown). As taurine itself has positive inotropic effects, adding it to the diet may be beneficial for long-term heart disease treatment, regardless of deficiency.

- **L-carnitine:** L-carnitine is abundant in skeletal and cardiac muscle and is essential for fatty acid metabolism and energy production. Some large breed DCMs have been linked to L-carnitine deficiency. L-carnitine is an **antioxidant and anti-steatotic agent** that prevents cell damage from oxygen free radicals. Supplementation with L-carnitine is needed in cases of heart failure or liver failure where ischemic damage or fat accumulation increases free radical production. While adverse effects of excessive L-carnitine are rare, the high cost can be a burden for owners.
- **Arginine:** Unlike other animal species, dogs and cats do not produce sufficient arginine endogenously, so it **must be included in the diet**. Arginine is a precursor to nitric oxide, a substance involved in **vascular smooth muscle relaxation (vasodilation)** and related to platelet function and inflammatory responses. Recent studies in humans showed a decrease in pulmonary artery pressure with L-arginine administration for pulmonary hypertension. Similar experiments in dogs did not confirm this effect. However, the administration method used in these dog studies resulted in low L-arginine bioavailability (citrulline conversion rate), suggesting that a change in administration method could benefit dogs with pulmonary hypertension or heart failure. Therefore, **L-arginine should be administered at 100 mg/kg, q8h.**

Potassium

Long-term administration of **furosemide (a diuretic) can cause hypokalemia** (especially in cats) due to increased renal potassium excretion. Hypokalemia can reduce appetite and vitality and increase susceptibility to digitalis toxicity. When ACEi are administered for heart disease treatment, they inhibit potassium excretion in the kidneys, which can cause **hyperkalemia**. Therefore, potassium supplementation should be avoided in animals receiving ACEi alone. However, in clinical practice, ACEi are rarely given alone and are usually administered with diuretics like furosemide. The increased potassium excretion caused by furosemide often **offsets the hyperkalemia induced by ACEi**, so the risk of ACEi-induced hyperkalemia is minimal; instead, treating hypokalemia is more important.

Commercial diets vary in potassium content, even within sodium-restricted diets, ranging from <150 mg/100 kcal to >300 mg/100 kcal. AAFCO guidelines recommend 170 mg/100 kcal for dogs and 150 mg/100 kcal for cats. If a dog with heart disease has hyperkalemia, a low-potassium diet should be chosen. Conversely, if the animal is hypokalemic, a high-potassium diet or oral potassium supplements should be given. Information on potassium content in diets should be obtained from the manufacturer. Feeding potassium-rich foods like bananas as treats can also be an option.

Vitamins

In humans, thiamine deficiency has been reported to cause cardiomyopathy. Although research on such vitamins in small animals is limited, recent studies suggest that Studies showed low vitamin B12 in feline cardiomyopathy or ateriothromboembolism and vitamin B12 level is negatively correlated with left atrial size. However, the evidence of B12 supplementation to reduce the incidence of cardiomyopathy is lacking. Studies in rats also show that Vitamin B12 deficiency increases myocardial hypertrophy and myocardial fibrin deposition, suggesting its usefulness as a dietary additive. **Supplementation with water-soluble vitamins is essential** in patients receiving long-term diuretic therapy, as these vitamins are more heavily excreted in urine.

Antioxidants

In humans, the therapeutic effects of antioxidants on heart disease have recently gained significant attention. **Reactive oxygen species**, generated through oxygen metabolism, are known to increase cell membrane vulnerability through peroxidation. Furthermore, an imbalance between endogenous oxidative waste and antioxidants has been reported as an aggravating factor in various diseases. Representative antioxidants include superoxide dismutase, catalase, glutathione peroxidase, Vitamin C, Vitamin E, and glutathione. Recent research indicates that **Vitamin E levels decrease in dogs as cardiomyopathy progresses.** Therefore, **antioxidant administration can aid in the treatment of heart disease.**

- **Coenzyme Q10 (Co-Q10)** is a coenzyme involved in energy production and a powerful antioxidant. While anecdotal reports suggest that Co-Q10 supplementation is effective in treating DCM in dogs, more systematically designed clinical trials are needed to verify its effects. The hypothesis that Co-Q10 deficiency is linked to myocardial disease also exists. Due to its potent antioxidant effects, Co-Q10 is actively recommended for patients with terminal cancer and heart failure in humans. The anectodal recommended dosage for dogs is **30 mg PO q12h** (up to 90 mg PO q12h can be administered in large dogs). Recent studies showed a daily CoQ10 dose of 200 mg was sufficient to achieve at least a 3-fold increase in plasma CoQ10 concentration and may be used in CoQ10 supplementation studies involving dogs with CHF due to MMVD.

2.3 Treatment Strategies for Heart Failure (American College of Veterinary Internal Medicine Guidelines)

The classification of heart failure progression in dogs and cats previously followed human heart failure staging, categorized into four stages, known as the Modified New York Heart Association (NYHA) Classification (Table 2-1). However, this classification often did not align well with dogs and cats, leading to the use of the International Small Animal Cardiac Health Council (ISACHC) classification for heart failure staging in small animal practice. The ISACHC classification has three main stages with five sub-stages (Table 2-2). More

Table 2-1 ▶ Modified New York Heart Association (NYHA) Classification

- **Class I:** Asymptomatic; heart murmur present but no cardiomegaly on radiography or echocardiography.
- **Class II:** Asymptomatic; heart murmur present and cardiomegaly on radiography or echocardiography, but no signs of pulmonary edema.
- **Class III:** Mild to moderate heart failure; heart murmur present, cardiomegaly on radiography or echocardiography, and signs of pulmonary edema on radiography.
- **Class IV:** Severe heart failure; heart murmur present, cardiomegaly on radiography or echocardiography, accompanied by severe dyspnea and pulmonary edema.

Table 2-2 ▶ International Small Animal Cardiac Health (ISACHC) Classification

- **Class Ia:** Asymptomatic; cardiac disease present but no clinical signs and no cardiomegaly on radiography or echocardiography.
- **Class Ib:** Asymptomatic; cardiac disease present but no clinical signs, and cardiomegaly on radiography or echocardiography (e.g., left atrial enlargement).
- **Class II:** Mild to moderate heart failure; cardiac disease present, clinical signs appear during exercise or excitement, and cardiomegaly on radiography or echocardiography, but not severe heart failure.
- **Class IIIa:** Severe heart failure; cardiac disease present, clinical signs like exercise intolerance at rest, and cardiomegaly on radiography or echocardiography, but treatable as an outpatient.
- **Class IIIb:** Refractory heart failure; cardiac disease present, clinical signs like exercise intolerance at rest, cardiomegaly on radiography or echocardiography, and severe enough to require hospitalization for intensive care.

recently, the American College of Veterinary Internal Medicine (ACVIM) has proposed a more detailed and specific classification system for heart failure stages (Table 2-3).

All these classification methods are based on clinical signs, and their subjective nature carries a risk of misjudgment depending on the owner's tendencies. Therefore, the 2019 ACVIM MMVD consensus statement uses a more objective classification method based on echocardiographic findings. However, even this objective assessment cannot overcome individual variations in symptom presentation, so the final determination must integrate objective cardiac examination data with the patient's clinical signs. Furthermore, sequential examinations at regular intervals and an integrated assessment of treatment responses can provide a more accurate prediction of the patient's heart failure stage.

Ideally, medical management of MMVD aims to stop the progression of valvular degeneration or improve valvular function. To date, no method has been found to prevent or stop valvular degeneration. However, surgical techniques or interventional therapies have been

Table 2-3 ▸ American College of Veterinary Internal Medicine (ACVIM) Guidelines

- **Stage A:** Individuals at high risk of developing heart failure (e.g., breeds prone to the condition like CKCS).
- **Stage B[1]:** Asymptomatic individuals with structural heart disease but no clinical signs related to heart failure.
  - **Stage B1:** Heart murmur present, but no evidence of cardiac remodeling (no cardiomegaly observed on radiography or ultrasound, or remodeling present but not severe enough to require treatment).
  - **Stage B2:** Hemodynamically significant and chronic mitral regurgitation observed on echocardiography or radiography, with evidence of left atrial/left ventricular enlargement. Drug therapy can delay HF progression. Ideally, B2 must fulfill all the following criteria(must consider breed variation): 1. murmur louder than 3/6. 2. VHS≥ 10.5. 3. LA/Ao≥1.6. 4. LVIDDN ≥1.7.
- **Stage C[2]:** Individuals with current or past clinical signs of HF due to MMVD (Myxomatous Mitral Valve Disease).
- **Stage D[3]:** End–stage MMVD patients with refractory heart failure that is not managed by standard heart failure therapies.

1) Patients with MR–related murmur but who have never experienced clinical signs of HF. The biggest difference from the 2009 guidelines is that the B2 group should initiate treatment to delay the onset of HF–related clinical signs.
2) Separated into emergency inpatient and outpatient treatment protocols. Some patients presenting with initial clinical signs may require more aggressive treatment similar to Stage D patients (e.g., afterload reducers and temporary patient support).
3) These patients require more aggressive management, and improvement without surgical correction is unlikely. This stage is also divided into emergency inpatient treatment and home outpatient treatment.

developed to improve valvular function. **Surgical mitral valve repair** has shown a good prognosis. However, such surgeries are only available in a limited number of regions, and are often not feasible for most dogs and cats due to economic or technical limitations.

Conversely, **atrial decompression**, which involves creating a hole in the atrial septum to shunt the pressure load from the left atrium to the right atrium, is gaining interest as a new treatment for patients with recurrent pulmonary edema. The author observed a sudden and significant improvement in clinical signs in a patient with end-stage MMVD and recurrent pulmonary edema without specific intervention. During follow-up, an atrial septal defect was identified. While this patient's case involved the rupture of the septum primum, unable to withstand severe left atrial enlargement, the sudden improvement led the author to hypothesize that **artificial atrial septotomy could help chronic symptom improvement in patients.** When this procedure was introduced into clinical practice, many end-stage heart failure patients (with recurrent pulmonary edema) showed improved clinical signs and extended survival. This has been reported by Dr. Allen and colleagues at VCA Animal Hospitals in the US. Unexpectedly, due to the excellent distensibility of the right atrium, side effects such as pulmonary hypertension did not occur despite the artificial left-to-right shunt. This interventional method should be **limited to end-stage MMVD patients with recurrent pulmonary edema** and performed only by experienced specialists (refer to Case 3). Puncturing the atrial wall instead of the septum can lead to severe hemorrhage and sudden death.

Therefore, the **pharmacological management of MMVD patients primarily focuses on improving quality of life**: extending the asymptomatic period, alleviating clinical signs, improving survival time, and maintaining weight. MMVD management is implemented by classifying patients into the following four categories: (1) Dogs with no CHF symptoms (ACVIM Stage B, ISACHC I), (2) Dogs with mild-to-moderate CHF symptoms (ACVIM Stage C, ISACHC II), (3) Dogs with recurrent CHF (ACVIM Stage C or D, ISACHC IIIa), and (4) Dogs with severe/critical CHF symptoms (ACVIM Stage C or D, ISACHC IIIb).

The following is a summary of the **2019 ACVIM (American College of Veterinary Internal Medicine) treatment guidelines for MMVD patients**:

- These guidelines are a revision of the 2009 guidelines.
- These guidelines adopt content from the American College of Cardiology guidelines, indicating the **strength of recommendation** for each. Additionally, the **strength of evidence** for each recommendation is provided.
- Recommendations are categorized from **Class I (therapeutic benefit far outweighs risk)** to **Class IV (risk outweighs therapeutic benefit)**.
- The **strength of evidence** for each recommendation ranges from **strong (based on results of one or more RCTs - Randomized Clinical Trials)** to **weak (based on uncontrolled clinical studies, case studies)** or **expert opinion (based on common sense, experience, or research in other animals, with weak academic evidence)**.

Strength of Recommendation

- **Class I : Strongly Recommended** (Benefit $\gg$ Risk)
- **Class II : Moderately Recommended**
  - **Class II a:** Benefit $\gg$ Risk
  - **Class II b:** Benefit $>$ Risk
- **Class III: Benefit and Risk are approximately equal** (Benefit = Risk)
- **Class IV : Risk outweighs Benefit** (Benefit $<$ Risk), Not Recommended

Strength of Evidence

- **Strong:** Based on results from one or more RCTs (Randomized Clinical Trials).
- **Moderate:** Based on well-designed observational studies and meta-analyses (retrospective studies).
- **Weak:** Based on uncontrolled clinical studies, case studies.
- **Expert Opinion:** Based on common sense, experience, research in other animals; weak academic evidence.

Diagnosis and Treatment Guidelines

Stage A (Patients at high risk of developing heart failure)

Diagnostic Recommendations:

✓ Regular annual check-ups are recommended for high-risk breeds (e.g., CKCS, Dachshund, Toy/Miniature Poodle Maltese) at primary care veterinary clinics (Class I, Expert Opinion).

✓ Active participation in screening programs hosted by dog shows or kennel clubs is recommended (for owners of high-prevalence breeds like CKCS) (Class I, Expert Opinion).

Treatment Recommendations:

✓ No recommended drug therapy (Class I, Expert Opinion).

✓ No recommended diet (Class I, Expert Opinion).

✓ Breeding dogs with echocardiographic findings or heart murmurs consistent with MMVD should not be used for breeding (Class I, Moderate).

Stage B (Patients with heart disease but no symptoms)

Diagnostic Recommendations:

✓ MMVD is often identified during routine health check-ups via auscultation. In asymptomatic MMVD patients, it is advisable to obtain thoracic radiographs to establish baseline data. Since MMVD patients often have co-occurring tracheobronchial disease, having baseline radiographic data makes it easier to differentiate between cardiac and non-cardiac causes of cough when clinical signs appear later (Class I, Expert Opinion).

✓ Echocardiography, including Doppler, is recommended to confirm the cause of the murmur, assess the extent of cardiac chamber enlargement, and identify co-existing conditions. It is beneficial to evaluate for co-existing pulmonary hypertension or elevated left atrial pressure. As mild left atrial or ventricular enlargement can be difficult to detect, it is crucial to compare findings using breed-specific reference ranges (Class I, Moderate).

✓ Comparing recent echocardiographic measurements such as LA/Ao and LV/Ao ratios in short-axis views and LA/LV ratios (refer to the diagnostic section) can more effectively detect cardiac enlargement in MMVD patients (Class I, Strong).

✓ When relying solely on thoracic radiographs to assess cardiomegaly, clinicians should be aware that VHS (Vertebral Heart Score) criteria can vary significantly with chest shape and breed-specific reference values. Vertebral left atrial size (VLAS) can be helpful (Class I, Moderate).

Additional Classification Recommendations for Group B:

✓ **Stage B1:** Asymptomatic patients for whom drug therapy does not delay the onset of heart failure; they have less or no cardiac enlargement compared to B2. Radiographic or echocardiographic findings show normal LA or LV diameter, or atrial/ventricular enlargement that does not categorize them into the B2 group (where drug therapy delays heart failure onset).

✓ **Stage B2:** Asymptomatic patients with sufficient cardiac enlargement (LA/LV) based on clinical trials, where drug therapy can delay the onset of clinical signs.

- Drug therapy is helpful in inhibiting heart failure progression in Stage B2 (Class I, Strong).
  - Heart murmur intensity > 3/6.
  - LA/Ao ≥ 1.6 (short-axis view).
  - LVIDd normalized to body weight (LVIDdN) ≥ 1.7 [LVIDd(cm)/BW(kg)0.294] = LVIDdN.
  - Thoracic radiographic VHS ≥ 10.5, considering breed.
- Ideally, all these criteria should be met before initiating treatment.
- Echocardiographic criteria among these are much more accurate than others and must be met.
- Radiographic criteria specific to Stage B2 have not yet been published. If echocardiography is unavailable, a VHS ≥ 11.5 clearly indicates cardiomegaly, and a pattern of increasing size on serial radiographs also indicates cardiomegaly and can substitute for echocardiography (Class I, Expert Opinion).
- A new radiographic indicator, VLAS, can predict left atrial size. While further research is needed, VLAS ≥ 3 suggests a high probability of Stage B2 MMVD (Class I, Moderate).

Treatment and Monitoring Recommendations for Group B:

✓ B1 Group:

- No drug therapy or diet recommended (Class I, Expert Opinion).

- Re-examination (echocardiography) every 6-12 months is recommended (more frequent follow-ups for large breeds) (Class I, Expert Opinion).

✓ B2 Group:

- **Pimobendan 0.25-0.3 mg/kg PO q12h** is recommended (Class I, Strong).
- **Dietary management is recommended.** Mild sodium restriction and adequate protein and calorie intake should be maintained to optimize BCS (Class IIa, Weak).
- 5/10 panelists recommended ACEi prescription (in countries where it is inexpensive) (Class IIa, Weak). Clinical trial results on ACEi efficacy are mixed.
- **β-blockers are not recommended**, regardless of cardiomegaly. Studies to prove the efficacy of β-blockers have not yet found specific benefits (Class III, Weak).
- **Spironolactone is also not recommended.** Large-scale clinical trials proving its efficacy have not been published yet, but preliminary studies have shown positive efficacy (Class IIb, Expert Opinion).
- Other drugs are generally not recommended. Some experts recommended β-blockers or amlodipine for advanced B2 stage patients, but further investigation into efficacy and safety is needed (Class III, Expert Opinion).
- If tracheal compression due to cardiomegaly is causing a cough, antitussives can be used for B2 stage patients (Class IIa, Expert Opinion).
- In regions where surgical correction is available, some experts recommended surgical treatment if the owner can afford the cost (Class IIa, Moderate).

Stage C (Patients with clinical signs of heart disease)

- Stage C refers to patients with MMVD severe enough to cause clinical signs of HF. This group includes patients who have experienced clinical signs of heart failure and respond to standard heart failure treatments. Even if clinical signs resolve with drug therapy, these patients remain classified as Stage C. Exceptionally, patients whose valve has been surgically corrected may be reclassified to Stage B. Although all Stage C patients respond to standard treatment, some critical Stage C patients may require more intensive heart failure treatment (sometimes applying Stage D patient treatment protocols). For acute heart failure patients in Stages C and D, the focus should be on controlling hemodynamic status and improving tissue oxygen delivery. For most patients where surgical correction is not feasible, the goals of chronic management are to improve hemodynamic parameters, slow heart failure progression, extend survival time, alleviate heart failure clinical signs, improve exercise capacity, maintain body weight, and improve quality of life.

Diagnostic Recommendations for Stage C:

✓ Breed, history, and physical examination should be performed prior to extensive testing. For example, an overweight dog without a history of weight loss is less likely to have clinical signs secondary to MMVD. Also, patients with severe sinus arrhythmia and a relatively slow heart rate are less likely to have a cardiac-induced cough compared to those with sinus tachycardia or normal sinus rhythm (Class I, Expert Opinion).

✓ Typical Stage C patients exhibit left heart failure clinical signs and a history of (tachypnea, restlessness, dyspnea, cough). Since breeds prone to MMVD also have a relatively high prevalence of chronic tracheobronchial disease, a cough in a dog with a left apical regurgitant murmur does not necessarily mean the symptom is caused by heart failure. Thoracic radiography, echocardiography, and basic laboratory tests should be performed to obtain baseline data. Renal dysfunction is a common co-morbidity in this patient group (Class I, Expert Opinion).

✓ Echocardiography, including Doppler examination, is useful for differentiating Stage C from D. These examinations can confirm valvular degeneration, quantify the degree of cardiomegaly and LV function, and measure left ventricular filling pressure. They can also identify co-morbidities (e.g., pericardial effusion, atrial rupture, or tumor origin) or complications (e.g., pulmonary hypertension/acquired atrial septal defect) due to chronic valvular insufficiency. A low mitral E-peak velocity strongly argues against a diagnosis of left heart failure, and most C and D Stage patients exhibit high E-peak velocities. In patients with clinical signs of pulmonary hypertension (e.g., exertional fatigue, collapse/syncope, ascites), Doppler examination can assess the presence and severity of pulmonary hypertension.

✓ **Serum NT-proBNP** is very helpful in differentiating between cardiac and non-cardiac causes of clinical signs. The group with heart failure-induced clinical signs has higher NT-proBNP concentrations than the group with respiratory disease-induced clinical signs. While the diagnostic value of a single NT-proBNP concentration has not been fully investigated, if a patient with cough, dyspnea, or exercise intolerance has a normal or near-normal NT-proBNP concentration, the clinical signs are unlikely to be due to heart failure (Class I, Moderate).

✓ Most MMVD patients are middle-aged or elderly, so blood pressure, CBC, biochemistry profile, and urinalysis data are necessary (especially for symptomatic MMVD patients requiring treatment) (Class I, Expert Opinion).

Recommendations for Stage C Acute Heart Failure Patients (Inpatient Treatment):

✓ **Furosemide 2 mg/kg IV (IM)** followed by 2 mg/kg IV/IM every hour or a total of 8 mg/kg over 4 hours until respiration stabilizes (Class I, Expert Opinion).

✓ For patients with critical pulmonary edema unresponsive to bolus furosemide (if respiratory rate does not improve after 2 hours), **furosemide CRI at 0.66-1 mg/kg/hr** (after a 2 mg/kg bolus) (Class IIa, Weak).

- ✓ Once diuresis has begun, allow the patient free access to water (Class I, Expert Opinion).
- ✓ **Pimobendan 0.25-0.3 mg/kg PO.** The use of pimobendan in acute patients is based on hemodynamic experimental results (some expert experience). IV pimobendan is available in some countries (Class I, Weak).
- ✓ **Oxygen supplementation** (via a humidity and temperature-controlled oxygen cage, incubator, or nasal oxygen cannula) (Class I, Expert Opinion).
- ✓ **Treat dyspnea-related sedation-excitement (anxiety).** Narcotics and tranquilizers are primarily used, so blood pressure and respiratory response must be closely monitored. While no specific recommendation exists, **butorphanol 0.2-0.25 mg/kg IM/IV** is most commonly used, and a combination of buprenorphine (0.0075-0.01 mg/kg) and acepromazine (0.01-0.03 mg/kg IV, IM, SC) or morphine and hydrocodone can be used (Class I, Expert Opinion).
- ✓ **Appropriate nursing care** (maintaining temperature and humidity; head on pillow, sternal position, etc.) (Class I, Expert Opinion).
- ✓ **Dobutamine (2.5-10 μg/kg/min CRI, starting at 2.5 μg and gradually increasing)**, administered along with the above treatments, can improve left ventricular function in patients unresponsive to those therapies. ECG monitoring is necessary to check for tachycardia or ectopic beats (Class I, Expert Opinion).
- ✓ **Sodium nitroprusside CRI (1-15 μg/kg/min) for up to 48 hours.** This method is useful for treating unresponsive pulmonary edema. If this drug is unavailable, oral arterial vasodilators like **hydralazine or amlodipine** can be used (Class I, Weak).
- ✓ **ACEi (enalapril or benazepril) 0.5 mg/kg PO q12h.** Although usually used for chronic treatment, some experts also use ACEi in acute patients. When used with furosemide and pimobendan, its efficacy and safety are somewhat unclear. However, PCWP (pulmonary capillary wedge pressure) was significantly improved when ACEi was co-administered compared to furosemide alone (Class IIb, Weak).
- ✓ **Nitroglycerin ointment, approximately 0.5 inch/10kg**, applied within 24-36 hours of hospitalization. Some experts recommend a 12-hour on/12-hour off protocol, while others did not use this method (Class IIb, Weak).

Recommendations for Stage C Chronic Heart Failure Patients (Home Care - Outpatient Treatment):

- ✓ **Oral furosemide** (typically 2 mg/kg q12h, dosage adjusted to patient comfort). Some experts substitute torsemide at 1/10-1/20 the furosemide dose (5-10% of furosemide dose) (0.1-0.3 mg/kg q24h) for patients not managed by furosemide (Class I, Moderate).
- ✓ Chronic PO furosemide dosages > 8 mg/kg q24h (or the equipotent torsemide dosage)

needed to maintain patient comfort in the face of appropriate dosages of **pimobendan, an ACEi, and spironolactone,** indicate disease progression to Stage D (Class I, LOE weak). Causes of diuretic resistance should be considered: poor compliance, high sodium intake, slow absorption (e.g., gut edema), impaired renal tubular secretion (e.g., CKD, aging, NSAID), hypoproteinemia, hypotension, nephron remodeling, neurohormonal activation (Class I, Weak).

- ✓ Check **serum creatinine, BUN, and electrolyte levels** within 3-14 days of starting furosemide treatment (Class I, Weak).
- ✓ Initiate or continue **ACEi therapy (0.5 mg/kg PO q12h)**. Check creatinine, BUN, and electrolyte levels within 3-14 days of starting ACEi. Suspect AKI if creatinine levels rise more than 30% from baseline (Class I, Weak).
- ✓ **Spironolactone (2 mg/kg PO q12-24h)**; aldosterone antagonism appears to be beneficial in treatment (Class I, Moderate).
- ✓ Continue **pimobendan** administration (Class I, Strong).
- ✓ Most experts did not recommend initiating **β-blockers** in situations where congestive heart failure symptoms had already manifested (Class IV, Weak).
- ✓ No experts used **nitroglycerin** for chronic Stage C patients (Class III, Expert Opinion).
- ✓ **Systematic home monitoring** (body weight, appetite, respiratory rate, heart rate) (Class I, Moderate).
- ✓ In centers with high surgical success rates, **mitral valve repair** can benefit patients (Class I, Moderate).
- ✓ In cases complicated by **atrial fibrillation**, **diltiazem** (often with digoxin) is recommended to control ventricular heart rate. Various diltiazem formulations exist, but start with a moderate dose and titrate to the amount that controls heart rate (Holter monitor needed). Maintain heart rate below 125 beats/min or within breed-specific ideal ranges (Class I, Moderate).
- ✓ **Digoxin 0.0025-0.005 mg/kg PO q12h** (maintaining plasma concentration at 0.8-1.5 ng/mL) can be added to slow ventricular heart rate in patients with persistent atrial fibrillation (generally co-administered with diltiazem). Digoxin can cause various signs of toxicity (Class IIb, Moderate).
- ✓ If a patient was receiving β-blockers before reaching Stage C, continued use of β-blockers was recommended even after progression to Stage C; some recommended reducing the dose (due to bradycardia, hypotension, hypothermia) (Class IIb, Expert Opinion).
- ✓ Some experts used **antitussives** (Class IIa, Expert Opinion).
- ✓ Some experts argued that **bronchodilators** are sometimes useful in some patients (Class IIb, Expert Opinion).

Dietary Recommendations for Stage C Patients:

- **Cardiac cachexia** refers to muscle loss or lean body mass loss associated with heart failure. Cardiac cachexia is a significant negative prognostic factor and is easier to prevent than to treat (Class I, Moderate).
- To minimize weight loss due to congestive heart failure, Stage C patients should maintain a **caloric intake of 60 kcal/kg**. Simple dietary modifications such as warming food, mixing dry and wet food, or offering various foods can improve appetite (Class I, Moderate).
- If anorexia occurs, identify the cause and correct it (Class I, Expert Opinion).
- Carefully record **BCS and body weight**, and determine the cause of weight changes (Class I, Expert Opinion).
- Encourage **adequate protein intake** and avoid low-protein diets for CKD patients (unless severe co-morbidities exist) (Class I, Moderate).
- **Moderately restrict sodium (Na) intake** from all foods, treats, and supplements. Avoid processed or salted foods (Class I, Moderate).
- **Monitor electrolytes** and supplement with potassium-containing supplements or natural foods. Some experts believe torsemide administration leads to more severe hypokalemia (Class I, Expert Opinion).
- Hyperkalemia is rare in patients receiving diuretics along with ACEi and spironolactone. If hyperkalemia occurs, avoid high-potassium diets (Class I, Expert Opinion).
- It is advisable to **monitor serum Mg levels** (especially as heart failure progresses; in patients with arrhythmias). If hypomagnesemia is present, supplement Mg (Class IIa, Expert Opinion).
- Consider **omega-3 supplementation** (especially in patients with decreased appetite, muscle loss, or arrhythmias) (Class IIa, Moderate).

Stage D (Refractory heart failure patients unresponsive to standard treatment)

- Stage D heart failure patients are defined as Stage C patients who do not respond to standard heart failure treatment (furosemide $\leq$ 8 mg/kg/d or torsemide $\leq$ 0.8 mg/kg/d, pimobendan 0.25-0.3 mg/kg PO q12h, a standard dosage of ACEi and spironolactone 2.0 mg/kg/d). Diagnosis is therefore made using the same methods as for Stage C, and by confirming that the patient does not respond to the standard Stage C treatment protocol.

Diagnostic Recommendations for Stage D:

✓ The diagnostic recommendations for Stage C, combined with evidence of unresponsiveness to standard heart failure treatment, allow for a diagnosis of Stage D.

Recommendations for Stage D Acute Heart Failure Patients (Hospital based, inpatient treatment):

- ✓ If severe renal failure is not present (creatinine > 3 mg/dL), administer **furosemide 2 mg/kg IV bolus** followed by either additional bolus doses or a furosemide CRI at a dosage of 0.66-1mg/kg/hr, until respiration stabilizes (for a maximum of 4 hours) (Class I, Expert Opinion).
- ✓ For patients no longer responding to furosemide, **torsemide (0.1-0.2 mg/kg q12-24h or 5-10% of the current furosemide dose)** can be administered. Torsemide causes less RAAS activation than furosemide. Once diuresis begins, provide sufficient water for free drinking (Class I, Expert Opinion).
- ✓ If necessary, **thoracocentesis or abdominocentesis** to stabilize respiration. Depending on the patient's condition, consider mechanical ventilation until the drug's effects manifest, in addition to oxygen supplementation (Class I, Weak).
- ✓ If the patient can tolerate it, **arterial vasodilators (Na nitroprusside)** can be administered with or without inotropic support (dobutamine) to reduce afterload until the effects of orally administered drugs (e.g., pimobendan or pimobendan + hydralazine/amlodipine) appear (most experts recommend this). Both drugs can be started at 1.0 μg/kg/min and titrated every 15-30 minutes until effects are observed, up to a maximum of 10-15 μg/kg/min. Blood pressure and ECG should be continuously monitored (Class IIa, Weak).
- ✓ Arterial vasodilators such as **hydralazine (0.5-2.0 mg/kg PO, gradually increased until maximum effect)** or **amlodipine (0.05-0.1 mg/kg PO, maximal effect 3 hours after oral administration)** can be added to reduce afterload (Class I, Expert Opinion).
- ✓ In addition to continuation of core medications (pimobendan and ACEi), more vigorous afterload reduction (arterial vasodilation) is generally recommended. Continuous monitoring is needed to prevent severe hypotension (systolic blood pressure < 85 mmHg). At least 24-72 hours after these treatments, serum creatinine levels should be rechecked. Although arterial vasodilators lower blood pressure, in patients with severe mitral regurgitation, they reduce afterload, increasing blood flow into the aorta, thus blood pressure may not significantly drop (Class IIa, Expert Opinion).
- ✓ For patients with signs of pulmonary hypertension (e.g., exercise intolerance, syncope, cough), **sildenafil 1-2 mg/kg PO q8h** can be administered (Class I, Moderate). If a patient with left heart failure has jugular distention or ascites, it suggests pulmonary hypertension, requiring immediate confirmation and sildenafil may be beneficial for such patients (Class IIa, Weak).
- ✓ **Pimobendan can be administered once more** (0.3 mg/kg, q8h). Some panelists state that if a patient presents with emergency pulmonary edema, they administer pimobendan once regardless of when the patient last received it at home (Class IIa, Expert Opinion).

✓ Some panelists recommended using **bronchodilators** in patients hospitalized for cardiogenic pulmonary edema (Class IIb, Expert Opinion).

Recommendations for Stage D Chronic Heart Failure Patients (Home Care-Outpatient Treatment):

✓ If renal failure does not occur, **furosemide or torsemide** should be increased to a level that prevents fluid retention (e.g., pulmonary edema or ascites). Drug-induced azotemia will worsen the patient's anorexia. To maximize drug efficacy, the following methods can be used: divide a twice-daily dose into three daily doses, or give one SC dose in between oral doses (Class IIa, Expert Opinion).

✓ For patients not managed by furosemide, **torsemide can be substituted at 1/10-1/20 the furosemide dose (5-10% of furosemide dose)** (maximum dose 0.6 mg/kg q12h) (Class I, Moderate).

✓ **Spironolactone (2 mg/kg PO q12-24h)**; aldosterone antagonism appears to be beneficial in treatment (Class I, Moderate).

✓ **Do not initiate β-blockers at this stage** unless for controlling atrial fibrillation (Class IV, Expert Opinion).

✓ Some panelists recommended adding **hydrochlorothiazide** (every other day). However, some panelists warned that this method significantly increases the risk of severe electrolyte disturbances and renal failure (Class IIb, Expert Opinion).

✓ **Pimobendan can be administered once more** (0.3 mg/kg, q8h). Higher doses may be administered depending on the situation (Class IIa, Expert Opinion).

✓ Adding arterial vasodilators such as **hydralazine (0.5-2.0 mg/kg PO, gradually increased until maximal effect)** or **amlodipine (0.05-0.1 mg/kg PO, maximal effect 3 hours after oral administration)** can be hemodynamically more beneficial and further reduce cough frequency.

✓ Unless specific side effects occur, patients with atrial fibrillation can be given **digoxin 0.0025-0.005 mg/kg PO q12h** (maintaining plasma concentration at 0.8-1.5 ng/mL) (Class IIb, Moderate).

✓ After stabilizing ventricular heart rate with digoxin or diltiazem, **β-blockers can be useful for reducing ventricular heart rate.** However, caution should be exercised regarding the negative inotropic effects of β-blockers on the ventricle (Class IIb, Expert Opinion).

✓ The majority of panelists recommended that if a patient is already on β-blockers, **do not discontinue them.** If symptoms are not improving with other medications (or if hypotension or bradycardia occurs), the drug dose should be reduced (Class IIb, Expert Opinion).

✓ Some experts recommended **antitussives** for uncontrolled chronic cough (Class IIa, Expert Opinion).

- ✓ Some experts recommended **bronchodilators** for uncontrolled chronic cough (Class IIb, Expert Opinion).

Dietary Recommendations for Stage D Patients:

- ✓ Same as for Stage C patients.
- ✓ If the patient has refractory fluid retention, **additional sodium restriction** is recommended as long as it does not worsen appetite or renal function (Class IIa, Expert Opinion).

Problems with the ACVIM Guidelines

1. Initiation of Treatment for Group B1 Patients

Most cardiologists generally agree that treatment should not be initiated for asymptomatic patients without cardiac enlargement (Stage B1). However, this criterion should be applied differently depending on the patient's situation. These guidelines are tailored to the circumstances in the US and Europe and do not fully represent the situation in countries like Korea or other Asian nations, where small, elderly breeds are overwhelmingly prevalent. For instance, small breeds such as Maltese and Shih Tzu have the highest incidence of MMVD in Asian countries, but are relatively less common in the US or Europe. Not only do these breeds have a high incidence, but if they develop MMVD, it usually progresses to heart failure, and the progression rate is faster compared to other breeds. However, no follow-up studies on drug treatment for asymptomatic patients in these high-incidence breeds have been conducted.

Furthermore, in Europe, where CKCS (Cavalier King Charles Spaniel) are overwhelmingly common, most research focuses on this breed. As a result, the actual research findings can be significantly different from what is observed in Korea and Taiwan. The discrepancies in the efficacy of ACEi seen in the SVEP and VETPROOF studies also stem from differences in the breed proportions within the study populations.

Our opinion is that even in Stage B1 patients without cardiac enlargement, the decision to administer ACEi preventatively should be based on the patient's breed and age. Empirically, Toy Poodles and Yorkshire Terriers are also prone to MMVD, but the disease progression is often slow, and many do not develop heart failure even at 15 years of age (in the case of Yorkshire Terriers, coughing is often due to tracheal collapse rather than heart failure). Therefore, in these high-incidence breeds, early ACEi administration is given preventatively.

The biggest problems with clinical trials in veterinary medicine are the small sample sizes, inconsistent diagnostic indicators, and population stratification issues where specific breeds are overrepresented. Consequently, there is a lack of well-designed clinical trials, and no proper clinical studies have been conducted specifically on Stage B1 patients. Follow-up studies on asymptomatic patients require long research periods and large study populations, and the research environment to conduct such studies is not yet established in the veterinary field. Therefore, in our practice, even for patients in the B1 group, we prescribe **ACEi** if the patient belongs to a high-incidence breed or shows a rapidly progressing pattern. Indeed, ACEi has several beneficial effects for cardiac patients, and long-term use typically does not cause significant side effects. The nephrotoxicity associated with ACEi is not caused by the drug itself but by its renal vasodilatory effect, so as long as adequate hydration is maintained and nephrotoxic drugs are not co-administered, long-term use does not induce nephrotoxicity.

2. Problems with the Diagnostic Guidelines for Group B2

Many cardiologists, including Professor Gordon from Texas A&M University, have raised questions about the ambiguous diagnostic guidelines for the B2 group. The B2 group consists of asymptomatic patients with cardiac enlargement. The main symptoms of heart failure are **exercise intolerance (easily tiring), cough, and dyspnea due to fluid retention**. Syncope is associated with arrhythmias or pulmonary hypertension, and these symptoms can occur even without severe heart failure. Among the three clinical signs mentioned above, exercise intolerance and cough are too ambiguous to serve as clear diagnostic guidelines, excluding dyspnea due to fluid retention.

While exercise intolerance is a crucial clinical sign in human heart failure patients, most MMVD patients are elderly and less active. Many heart failure patients may not exhibit this symptom even during walks. Furthermore, it's often difficult for owners to observe changes in exercise intolerance.

Regarding cough, recent textbooks and review articles emphasize that the tracheal compression due to cardiomegaly, as we once understood it, is **not the primary cause of cardiogenic cough**. Moreover, the consensus in the academic community is that pulmonary edema does not induce coughing because the alveoli and small bronchi hardly contain cough receptors. Therefore, the main causes of cough in heart failure patients are believed to be **tracheal collapse, chronic tracheobronchitis, and pulmonary hypertension**. Recent review articles thus point out that a cough in a cardiac patient cannot necessarily be considered a symptom of heart disease. This situation makes the diagnosis of the B2 group even more challenging.

The imaging criteria for diagnosing the B2 group are **LA/Ao $\geq$ 1.6 and LVIDdN $\geq$ 1.7 on echocardiography**. However, the LA/Ao criterion can vary depending on the measurement method, and LVIDdN, while attempting to exclude differences in LVIDd due to body weight, is not an accurate criterion as it doesn't account for different body condition scores (BCS) in patients of similar build (e.g., obese and thin patients of similar size may have different LVIDdN values). A **thoracic radiographic VHS $\geq$ 10.5** is not very helpful because it varies greatly with breed chest shape. Methods for measuring left atrial size like **VLAS or RLAS** are often impractical or ambiguous in actual patients, making them unreliable diagnostic criteria.

Furthermore, some patients exhibit clinical signs earlier, while others show them later, depending on individual variation. Also, depending on the owner's personality, even subtle clinical signs might be detected early or late. In reality, the echocardiographic diagnostic criteria for asymptomatic populations used in many drug trials vary widely, with LA/Ao ranging from 1.4 to 2.0, and the study results differ significantly depending on the median LA/Ao of these research populations.

The administration of ACEi and pimobendan in the B2 group is undeniably beneficial, as proven by guidelines and numerous prior studies. However, there's a concern that delaying more potent heart failure treatments like diuretics for asymptomatic B2 group patients who are masking symptoms could worsen their condition.

Professor Gordon's recent review article proposes an additional **B+ group** within the B group. The B+ group refers to patients who do not show clinical signs of fluid retention but have sufficient cardiac enlargement and increased left atrial pressure, suggesting that diuretic treatment could be beneficial. We consider **E/E' ≥ 9, %LA ≤ 30%, IVRT ≤ 45 ms, and E/IVRT ≥ 2.5** as indicators of increased left atrial pressure and initiate diuretic treatment depending on the patient. Appropriate early diuretic treatment reduces left atrial preload, preventing further cardiac enlargement, and in actual patients, it has been shown to extend survival and delay the onset of clinical signs.

The ACVIM guideline that recommends initiating diuretic treatment only when pulmonary edema or ascites occurs is **not suitable for the Asian context**. The reported survival rates for MMVD patients in Europe and the US are approximately 50% shorter than those of patients treated at our clinic and our colleagues' hospitals. Although it's unclear whether the published research results are based on general hospitals or large specialty hospitals with cardiologists, our treatment approach, which adjusts diuretic introduction and dosage based on echocardiographic criteria, demonstrates longer survival periods. Therefore, when selecting B2 group patients, it is advisable to **initiate diuretics early in patients with cardiac enlargement or increased left atrial pressure on imaging consistent with the C group (B+ group)**. According to Professor Gordon's review article, the following three criteria are suggested for differentiating the B+ group, which is on the verge of heart failure symptoms: 1) patients with significantly large cardiac enlargement on imaging, 2) patients with persistent cardiac enlargement on repeated examinations, and 3) patients maintaining high concentrations of cardiac biomarkers like NT-proBNP.

3. Problems with the Treatment Strategies for Groups C and D

Many heart failure patients in Group B can progress directly to Group D without passing through Group C. Early detection of pulmonary edema depends on the patient's compliance, the owner's observation skills, and the frequency of patient monitoring. The severity of symptoms can also vary depending on the occurrence of complications (e.g., chordal rupture, co-existing conditions causing fluid retention). Therefore, some patients in Group C may also require the aggressive acute treatment typically given to Group D.

The acute treatment protocols for Group C and Group D differ in the use of arterial vasodilators and fluid removal. In our practice, we do not make a significant distinction between acute treatment methods for these groups. Although the initial treatment may vary based on the patient's imaging findings, regardless of whether they are in Stage C or D, we immediately initiate **furosemide and oxygen therapy**, and if right heart failure is also present, we remove ascites. If the patient's systolic blood pressure is at or above normal levels (≥ 110 mmHg), we co-administer **sodium nitroprusside 2.5 μg/kg/min CRI**. If oral administration is possible, we give **pimobendan orally**. Empirically, intravenous pimobendan does not increase systolic blood pressure as much as dobutamine. Therefore, for hypotensive patients, we administer **dobutamine 5 μg/kg/min CRI**. If there's no response after a 2-4 mg/kg bolus of furosemide, we start a CRI at 0.66 mg/kg/hr. For pulmonary edema patients unresponsive to these methods, we may administer aminophylline or intravenous amlodipine. Therefore, we use a progressive treatment approach, moving from lower-stage treatments to higher-stage ones as appropriate for the patient's situation, rather than subcategorizing acute pulmonary edema patients into Group C and Group D.

The onset and severity of pulmonary edema do not always seem to correlate proportionally with the severity of cardiomegaly. The timing and severity of pulmonary edema seem to vary greatly depending on the patient's personality, complications, or co-existing conditions. The ACVIM guidelines fail to address these aspects.

4. Application of Pulmonary Hypertension Drugs like Sildenafil

According to the recently published ACVIM diagnostic and treatment guidelines for pulmonary hypertension, the use of **sildenafil is restricted in Class II (pulmonary venous hypertension)**. PDEi (phosphodiesterase inhibitors) like sildenafil expand pulmonary arteries, thereby increasing blood flow into the pulmonary veins. Ultimately, these drugs increase left atrial pressure, which can worsen clinical signs of pulmonary edema in patients with impending or existing pulmonary edema.

The ACVIM guidelines recommend the use of sildenafil for Stage C and D patients if they exhibit symptoms or imaging findings of pulmonary hypertension. This is a **very risky recommendation** for two main reasons: 1) The listed clinical signs of pulmonary hypertension are not significantly different from general heart failure symptoms, making it difficult for general veterinarians to differentiate them. 2) Imaging findings of pulmonary hypertension are often challenging to distinguish unless interpreted by a highly trained cardiologist. The most straightforward echocardiographic findings like TR (tricuspid regurgitation) or PR (pulmonic regurgitation) often do not appear even when pulmonary hypertension is present, and other echocardiographic parameters are not widely used. Therefore, the imaging diagnostic section of this book covers these aspects in more detail. It is important to know that administering **sildenafil** to patients without ascites or jugular distention can actually **worsen pulmonary edema or left heart failure symptoms**. Alternatively, it is advisable to consult a cardiologist to accurately assess the presence of pulmonary hypertension before deciding whether to prescribe sildenafil.

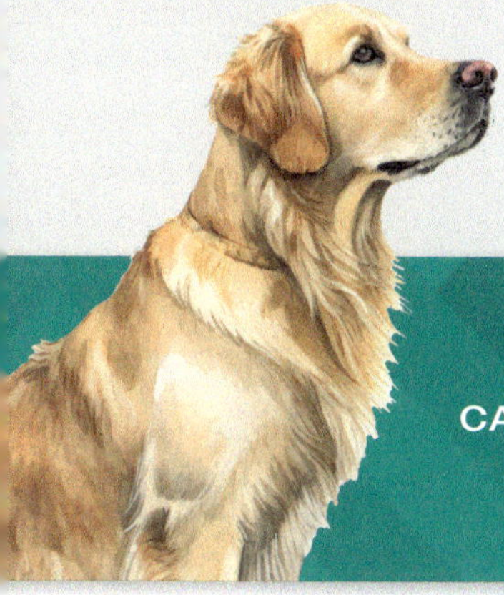

CHAPTER 03

Myxomatous Mitral Valve Disease

3.1 Introduction

Myxomatous mitral valve disease (MMVD) is the most common acquired heart disease in dogs. This condition is known by various names based on its clinical and pathological characteristics, including **mucoid valvular degeneration (MVD), chronic mitral valvular insufficiency (CMVI), valvular endocardiosis, myxomatous transformation, and degenerative valvular disease (DVD)**. However, in clinical practice, it is often simply referred to as **mitral regurgitation (MR)**. The mitral valve is unique among heart valves for having two leaflets and is located between the left atrium and the left ventricle. Its function is to prevent blood that has entered the left ventricle from flowing back into the left atrium during cardiac systole.

MMVD is a disease where myxomatous degenerative changes occur in the mitral valve leaflets, causing them to thicken and become nodular, preventing the valve from closing properly (Figure 3-1A). Histologically, MMVD is characterized by changes in the **intercellular matrix** of the valve leaflets, including the cellular components and chordae tendineae, with deformities occurring in both **collagen components and collagen fibrils** within the valve (Figure 3-1B, C). While the endothelial and subendothelial cells of the heart in MMVD patients may degenerate and hypertrophy, this does not increase the likelihood of thrombus formation in the left atrium or the risk of endocarditis. Ultimately, this progressive valve degeneration impairs the valve's primary function of opening and closing (insufficiency), increasing the volume of blood regurgitated into the atrium. This leads to atrial enlargement and ultimately causes volume overload in the ventricle, leading to **eccentric hypertrophy (dilation)** of the ventricle and progression to congestive heart failure.

The cause of MMVD in small animals has not yet been identified, but aging and genetic influences in some breeds have been reported. Specifically, in **Cavalier King Charles Span-**

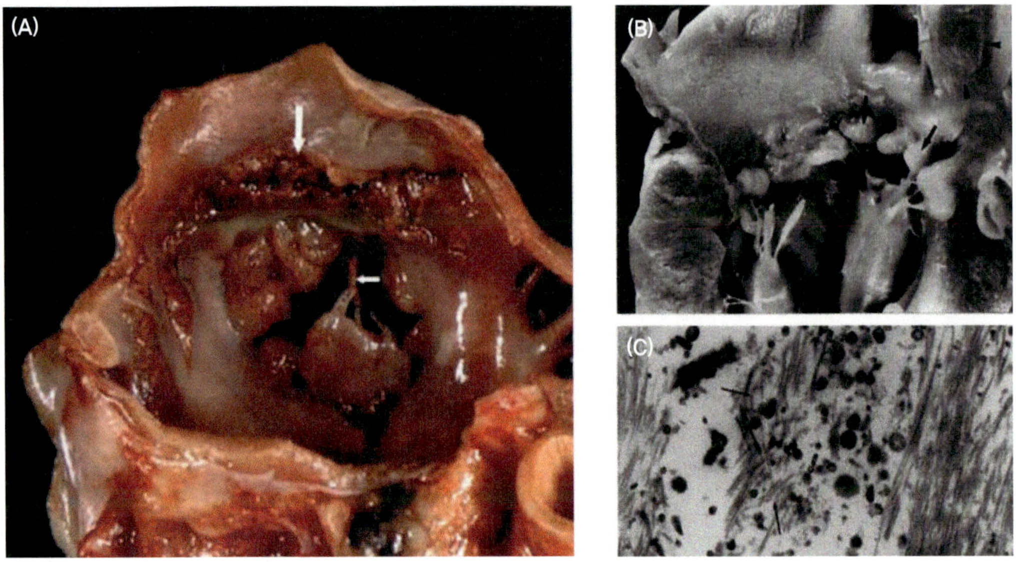

Figure 3-1 ▶ **Gross (A) and Electron Microscopic (B, C) Findings of Myxomatous mitral valve disease.** MMVD is a disease where myxomatous degenerative changes occur in the mitral valve leaflets, causing them to thicken and develop nodular changes, preventing proper valve closure. Histopathologically, it is characterized by changes in the intercellular matrix of the valve leaflets, including cardiac cellular components and chordae tendineae. Both collagen components and collagen fibrils within the valve leaflets undergo deformation.

iels and Dachshunds, it has been found to be inherited as an autosomal dominant trait, sharing very similar disease patterns and genetic characteristics with human mitral valve prolapse (Figure 3-2). However, this disease is said to exhibit a **polygenic trait** where the phenotype is determined by multiple genes, rather than a single-gene transmission pattern explained by Mendelian laws. Studies suggest that combinations of dams and sires with early-onset MMVD produce offspring with early-onset MMVD, while combinations of dogs with late-onset MMVD produce offspring that develop the disease later in life or do not express it at all. Additionally, male dogs are more susceptible to the disease, developing it at a younger age than female dogs. Recent research has identified genes related to early-onset MMVD in CKCS breeds at genetic loci on chromosomes 13 and 14.

While genetic factors play a major role in disease development, other factors such as exercise level, obesity, and dietary factors are also said to contribute, albeit minimally. Changes in the **mitogen receptors of fibroblast cell membranes** appear to play an important role in the pathogenesis of this disease. Metabolic, neuro-hormonal, or inflammatory mediators are believed to be deeply involved in the pathological progression of the valve, myocardial remodeling, and the development of ventricular dysfunction, but the exact cause has not yet been elucidated. Furthermore, The factors determining the progression of valve degeneration are not entirely unknown. Age, progressive heart enlargement (left atrium and left ventricle), increased transmitral E wave blood flow velocities, increased serum N-terminal pro-B-type natriuretic peptide(NT-proBNP) concentrations, and increases in resting

Figure 3-2 ▸ Myxomatous mitral valve disease in Cavalier King Charles Spaniels has been identified as an autosomal dominant inherited condition.

heart rate are at least moderately predictive of the rate of progression of MMVD. A recent study reported that over 85% of dogs older than 13 years showed varying degrees of valve degenerative changes at necropsy. These findings suggest that the incidence of this disease is closely related to age-related degenerative changes. However, not all cases of valve degeneration progress to heart failure, so further research should focus on identifying the factors that cause functional impairment in individuals showing such pathological changes.

Breeding programs to suppress disease incidence are being implemented for CKCS and Dachshunds in North America and Europe. These screening programs select candidates through auscultation and identify affected individuals through echocardiography. Such screening methods appear to be necessary for Maltese and Shih Tzu breeders in Korea as well.

3.2 Pathogenesis

MMVD involves degenerative changes in the mitral valve, leading to thickening and nodularization of the valve leaflets, resulting in insufficiency. The increased blood volume regurgitated into the left atrium due to mitral valve insufficiency induces left atrial enlargement. The regurgitated blood, along with the normal pulmonary venous return, re-enters the ventricle during diastole, causing **volume overload** in the ventricle. As a compensatory response of ventricular cells to volume overload, myocardial cells do not increase in number but individual cells lengthen, leading to the **dilation and stretching of the ventricle itself (eccentric hypertrophy)**. Such stretched myocardial cells, in the long term, cannot exert normal contractility and require a lot of energy, making them much more vulnerable to ischemic damage. These cells can cause conduction disorders, leading to arrhythmias or sudden death.

The left ventricle, dilated by eccentric hypertrophy, can initially compensate for impaired cardiac function through neurohormonal system activation (e.g., increased heart rate and increased systemic blood volume through sodium retention). However, gradually, contractility decreases due to increased circulating blood volume and increased vascular resistance. If the left ventricle can no longer maintain its compensatory response, the regurgitated blood volume further increases, leading to left atrial enlargement and a significant increase in left atrial end-diastolic pressure. Consequently, the left atrial pressure further increases due to blood regurgitating into the left atrium during ventricular systole, resulting in pulmonary venous congestion. When the hydrostatic pressure in the pulmonary vasculature rises due to this pulmonary venous congestion, **pulmonary edema** occurs, where fluid accumulates in the interstitium. This pulmonary edema is a characteristic symptom of left heart failure and is typically observed in most progressive MMVD patients.

While mild-to-moderate tricuspid valve myxomatous degeneration can cause tricuspid regurgitation, the right heart system has good compliance and tolerates volume increases well. However, the right ventricle is vulnerable to pressure increases because it pumps blood into the low-pressure pulmonary vasculature. Therefore, if left heart failure increases pulmonary venous pressure, leading to an increase in pulmonary arterial pressure, RV output can significantly decrease. This complication is called **Type II pulmonary hypertension**. Right ventricular enlargement due to pulmonary hypertension can cause right atrial enlargement, which in turn can lead to atrial tachyarrhythmias (e.g., atrial fibrillation and supraventricular tachycardia). Additionally, increased right atrial pressure can cause typical right heart failure symptoms such as **ascites, pericardial effusion, hepatomegaly, and splenomegaly**.

3.3 Clinical Findings

A key characteristic of early MMVD is a **systolic murmur**. In most dogs, this murmur is observed before clinical symptoms appear, and it often takes several years or more for clinical symptoms to manifest after the murmur is detected. Rarely, dogs with MMVD may not develop clinical signs of heart failure before succumbing to noncardiac disease. In humans, **exercise intolerance (easily fatiguing)** is often the first symptom of heart failure. However, since MMVD primarily affects elderly dogs with low activity levels, owners may not easily detect this symptom. Therefore, classifying patients based on the presence or absence of clinical symptoms in MMVD diagnosis always carries the risk of misdiagnosis. Thus, the common assertion in ACVIM guidelines and internal medicine textbooks that clinical symptoms of heart disease are not readily apparent in mild-to-moderate MMVD patients is somewhat problematic.

Especially considering recent research findings that the bronchioles and alveoli have very few cough receptors, the long-held belief that cough symptoms are caused by compression of the left main bronchus due to left atrial enlargement may be incorrect. A significant number of small, elderly dog breeds often have tracheal collapse or chronic tracheobronchitis, which can cause coughing. Therefore, it seems necessary to reconsider treating cardiogenic cough as a primary clinical symptom. However, the fact that diuretics or cardiac medications often help alleviate coughing in these patients suggests that cardiomegaly may be a contributing factor to coughing. However, furosemide has antiinflammatory and antitussive properties, meaning that a cough responding to diuretic therapy does not necessarily confirm a diagnosis of CHF. Therefore, strategies aimed at reducing cardiomegaly may alleviate coughing by reducing the component that stimulates cough receptors,rather than directly removing the cause of the cough.

Clinical symptoms of MMVD accompanied by heart murmur include **cough, exercise intolerance, dyspnea, anorexia, weight loss, syncope, and tachycardia**. Left heart failure due to mitral valve damage can cause nocturnal cough and severe pulmonary edema, leading to dyspnea. Tracheal compression due to cardiac enlargement typically causes a dry, harsh cough, but if accompanied by pulmonary edema, it can cause a wet cough. Furthermore, in very severe pulmonary edema, hemoptysis with modified transudate may be observed. While tracheal compression due to left atrial enlargement can cause coughing, it is difficult to differentiate from symptoms of chronic bronchial disease. Differentiation can be attempted with radiography and a trial of diuretics. In some cases, reduced cerebral blood flow due to **forward heart failure** can cause syncope. This occurs because cardiac output is insufficient to meet the body's physiological demands.

Syncope can be observed in some MMVD patients, primarily caused by **tachyarrhythmias or vagal hyperactivity (vasovagal syncope)**. Other causes of syncope include **tussive fainting** (occurring in relation to explosive coughing fits) or exercise in a state of **pulmonary hypertension**. Simple TR without pulmonary hypertension rarely causes clinical disease. However, if **Type II pulmonary hypertension** develops due to increased left atrial pressure, symptoms of **exercise intolerance, weakness, and syncope** may occur. In such patients, typical right heart failure symptoms (respiratory distress due to pleural effusion ± ascites, abdominal distension due to hepatomegaly or splenomegaly ± gastrointestinal symptoms such as diarrhea, vomiting, and anorexia) are commonly observed.

MMVD is primarily found in older dogs and occurs approximately **1.5 times more frequently in males than females**. Furthermore, according to studies evaluating valvular disease through postmortem evidence in dogs, severe valvular disease was found in approximately 58% of dogs over 9 years old, and if mild valvular degeneration is included, evidence of valvular degeneration is found in about 90% of dogs over 13 years old during postmortem examination. Regarding breed size, it predominantly affects **small breeds weighing less than 10 kg**, but sometimes large breeds can also be affected. A recent study found that, the frequency of cardiac-related death is low in large breed dogs with stage B1 MMVD. However, the risk of a negative outcome increases profoundly when large breed dogs advance from ACVIM stage B1 into stage B2 or C. Particularly, the **Cavalier King Charles Spaniel** breed is prone to developing the disease at a relatively young age (refer to Figure 3-2), but apart

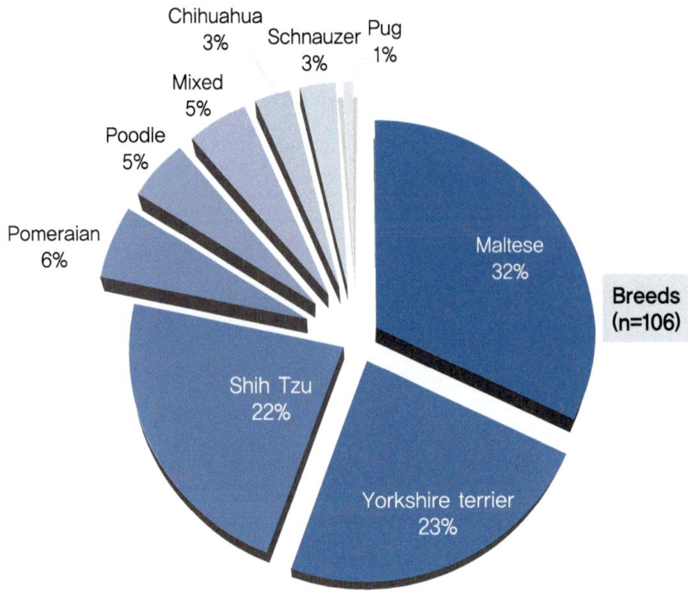

M-H Lee (2010) Retrospective study of 106 dogs with Myxomatous mitral valve disease. Master thesis, Kangwon Narional University, Korea.

Figure 3-3 ▸ Prevalence of myxomatous mitral valve disease (MMVD) in Korea. The three most common breeds affected by MMVD in Korea are Shih Tzu, Maltese, and Yorkshire terrier.

from cases in young puppies, the clinical signs are largely similar to those occurring in other small breeds.

Looking at the positional incidence of valve insufficiency, isolated mitral valve disease accounts for approximately 60%, isolated tricuspid valve disease for about 10%, and the remaining 30% involves insufficiency in both mitral and tricuspid valves. Regarding breed specificity, it commonly affects **Cavalier King Charles Spaniels, Dachshunds, Yorkshire Terriers, Chihuahuas, Miniature Schnauzers, and other small breeds**. In Korea, the incidence in **Maltese and Shih Tzus** is reported to be very high (Figure 3-3). This seems to be related to the cultural preference for small dog breeds in Korea. On the other hand, this disease is uncommon in cats, and even if it occurs, there are rarely clinical symptoms. In breeds like Maltese and Shih Tzu, the age of onset is early, and almost all cases progress to heart failure. In contrast, while the incidence is high in Toy Poodles and Yorkshire Terriers, the progression to heart failure is slow, and they are often mistakenly treated for heart failure due to coughing symptoms caused by co-occurring chronic bronchitis or tracheal collapse rather than the heart disease itself. Since not all patients with heart murmurs and echocardiographic evidence of mitral regurgitation require diuretic treatment for advanced MMVD, it is advisable to thoroughly investigate the co-occurrence of respiratory diseases in these breeds before initiating heart failure treatment. MMVD is a disease closely related to aging, so its incidence is higher in the elderly population. Furthermore, the elderly population often has more severe heart failure.

3.4 Diagnosis

Prior to diagnosing MMVD, it is crucial to differentiate it from conditions such as **tracheal collapse, chronic bronchitis, bronchiectasis, pulmonary fibrosis, pulmonary tumors, pneumonia, pharyngitis, heartworm disease, and dilated cardiomyopathy** (DCM).

After checking the patient's **vital signs** such as weight, temperature, blood pressure, respiratory rate, and heart rate, a thorough auscultation with a stethoscope should be performed. Generally, patients affected by MMVD will have a characteristic **systolic murmur** heard over the left apical region of the heart (Figure 3-4). In cases of mild disease, a faint **mid-systolic click** (Figure 3-4A) or a **brief soft systolic murmur** may be heard, and the patient's respiratory rate may increase. If the disease is advanced, the heart murmur will intensify, presenting as a loud. As the disease progresses, the murmur becomes harsher and louder, radiating to the right chest wall, which can make it difficult to differentiate between mitral and tricuspid insufficiency.

Generally, laboratory findings in MMVD patients are normal or may show mild elevations in kidney or liver values. These laboratory tests can help differentiate coughing caused by infectious respiratory diseases or respiratory tumors from coughing due to cardiac enlargement. Additionally, decreased cardiac output due to heart disease can occur, leading to reduced renal perfusion and sometimes causing **cardio-renal syndrome**. Consequently, various laboratory changes due to prerenal azotemia may be observed. Recently, several **cardiac biomarker tests** have been developed for dogs and cats to assess the progression of heart failure and differentiate between cardiac and respiratory diseases. Among these, **N-terminal probrain natriuretic peptide (NT-proBNP)** is known to be a very useful test for evaluating the progression of MMVD (Table 3-1).

During an electrocardiogram (ECG) examination, findings indicating left atrial enlargement, such as **P mitrale (wide or/and notched P wave)**, and signs of left ventricular hypertrophy, such as **tall & wide QRS**, are commonly observed (Figure 3-5). As mitral in-

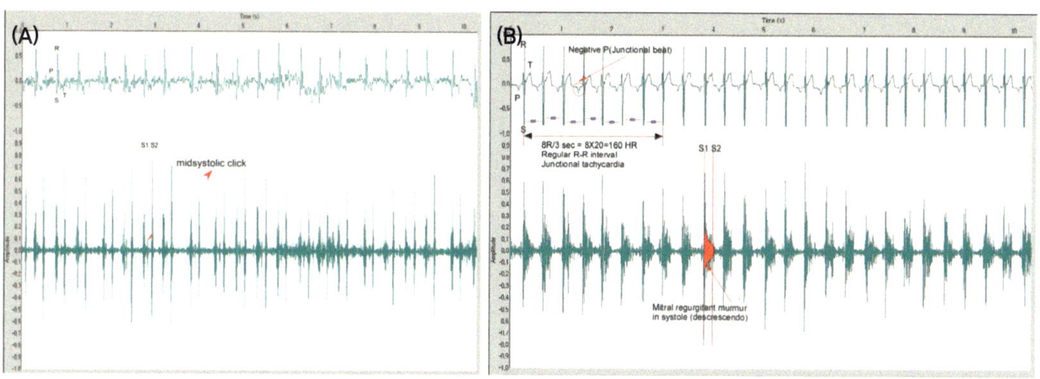

Figure 3-4 ▸ Phonocardiographic Findings of myxomatous mitral valve disease (MMVD). (A) Mid-systolic click auscultated in early MMVD patients. (B) Typical holosystolic plateau-shaped murmur. regurgitant murmur auscultated in MMVD patients with advanced disease.

Table 3-1 ▶ Interpretive Criteria for the Canine Cardiopet® proBNP Test

For Dogs Suspected of Heart Disease (Murmur or At-risk Breed)

- **NTproBNP <900 pmol/L**
  - NTproBNP concentration is not compatible with increased stretch and stress on the myocardium. Clinically significant heart disease is unlikely at this time.
- **NTproBNP ≥900 pmol/L**
  - In dogs with a murmur, NTproBNP concentration is compatible with increased stretch and stress on the myocardium. Clinically significant heart disease is likely at this time. Additional diagnostics are recommended to diagnose and assess severity of the disease.
  - For dogs <20 kg with mitral valve disease (MVD) and NTproBNP >1,500 pmol/L, Patient is at increased risk of heart failure in the coming 12 months. Thoracic radiographs and vertebral heart score (VHS), at a minimum, are required.

For Dogs with a Murmur and Clinical Signs Consistent with Cardiac Disease

- **NTproBNP <900 pmol/L**
  - The likelihood that clinical signs (i.e., respiratory and/or exercise intolerance) are due to heart failure is low.Consider other differentials to determine the cause of clinical signs.
- **NTproBNP 900–1,800 pmol/L**
  - There is increased stretch and stress on the myocardium at this time. However, results in this range do not allow reliable differentiation between clinical signs due to heart failure versus those from other causes.
  - Additional diagnostics are recommended.
- **NTproBNP >1,800 pmol/L**
  - There is evidence of increased stretch and stress on the myocardium. The likelihood that clinical signs (i.e., respiratory and/or exercise intolerance) are due to heart failure is high. Additional diagnostics are recommended to diagnose and assess severity of the disease.

- **For dogs with NTproBNP results 900 pmol/L or higher, additional diagnostics to assess cardiac health include thoracic radiographs, electrocardiogram and an echocardiogram. Echocardiography generally provides the most accurate information regarding cardiac structure and function**

sufficiency progresses, **tachycardia due to persistent or paroxysmal supraventricular premature beats** may occur. In large breeds, **atrial fibrillation** is often observed. However, in the early stages of the disease, the ECG is usually normal. If myocardial hypoxic damage occurs due to myocardial remodeling, signs of myocardial ischemia such as **ST-slurring** may be observed (Figure 3-5). Logically, if pulmonary hypertension is co-present, ECG findings of right heart enlargement such as RA enlargement (tall P-wave), RV enlargement pattern (deep S-wave), and right axis deviation of MEA may be observed. However, due to left heart system changes, these findings are often not observed.

On thoracic radiographs, left atrial and left ventricular enlargement are observed. Gen-

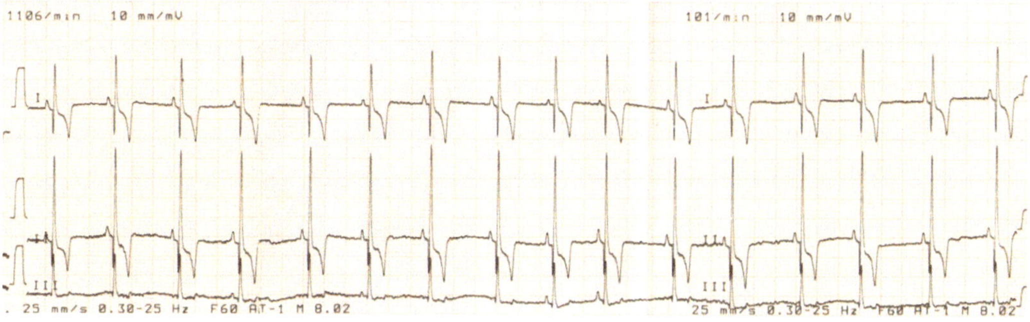

Figure 3-5 ▸ Electrocardiographic Findings of myxomatous mitral valve disease (MMVD). The patient is a progressive MMVD case showing typical left ventricular enlargement with an R-wave height of over 2.5 mV. Additionally, severe ST-slurring suggests increased ventricular wall stress and ischemic myocardial damage due to ventricular dilation.

erally, in MMVD patients, when left atrial and left ventricular dilation occurs, the heart appears elevated on lateral thoracic radiographs (increasing to more than 2/3 of the total thoracic height; Figure 3-6B). **Dorsal displacement of the trachea** and **compression and elevation of the left main stem bronchus** due to left atrial enlargement cause the left and right main bronchi to appear splayed (Figure 3-6B). The pulmonary veins are dilated due to increased hydrostatic pressure, and pulmonary edema is also observed (Figure 3-6). In the B2 stage, cardiac enlargement is observed without clinical symptoms, but as the disease progresses, **bronchial pattern and alveolar pattern** due to pulmonary parenchymal conges-

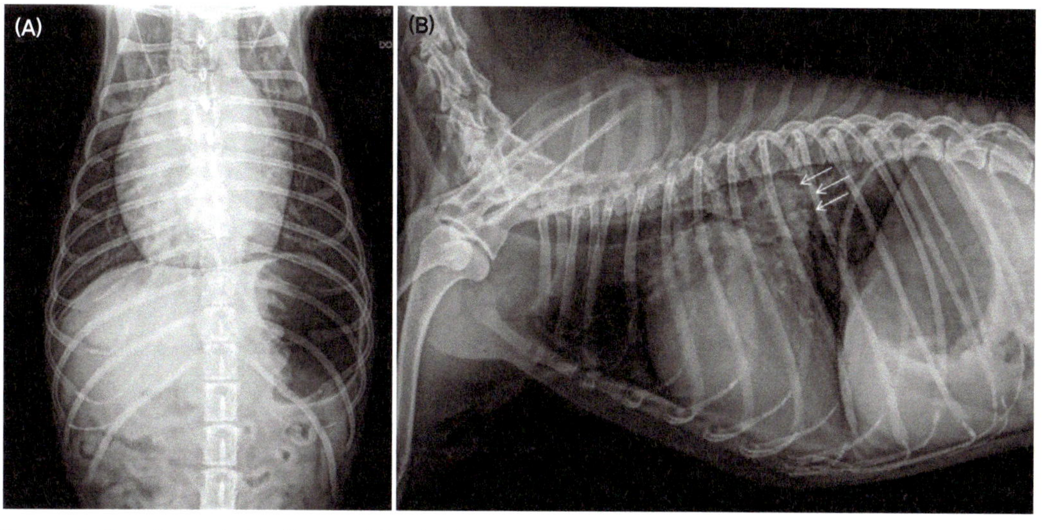

Figure 3-6 ▸ Thoracic Radiographic Findings of myxomatous mitral valve disease (MMVD). (A) Dorsoventral view – Cardiomegaly, bronchial pattern, and alveolar pattern due to congestion and edema in the pulmonary parenchyma are observed. (B) Lateral view – Increased cardiac height (increased to more than 2/3 of the total thoracic height), dorsal displacement of the trachea, and compression of the left main stem bronchus are observed. Most notably, marked left atrial enlargement is observed (arrow).

tion and edema are observed (Figure 3-6A). These findings suggest cardiogenic pulmonary edema, which typically has a perihilar distribution with a predilection for the right caudal lung lobe in dogs. **Air bronchograms** due to alveolar patterns can be observed, and such alveolar pulmonary edema indicates severe congestive heart failure in patients experiencing dyspnea. If **chordae tendineae rupture** occurs, severe pulmonary edema may be present even without signs of left cardiomegaly. If pulmonary hypertension is co-present, signs of pulmonary artery dilation along with right atrial and right ventricular enlargement are also observed.

Among cardiac examination methods, **ultrasound (echocardiography)** is a non-invasive method that can detect valve disease from its early stages. For evaluating MMVD, **2-dimensional (2D), M-mode imaging, and Doppler ultrasound** are commonly performed. When performing an ultrasound examination to evaluate the heart, a 2D examination is always performed first to identify basic abnormalities and measure reference values. Then, M-mode and Doppler ultrasound are performed to pursue diagnostic clues. However, the most important diagnostic evidence may not always be confirmed on ultrasound images, so ultrasound findings should ideally be evaluated in conjunction with the patient's clinical signs. Therefore, for an accurate echocardiographic diagnosis, a meticulous physical examination and history taking should be performed before evaluation.

There are various echocardiographic diagnostic criteria for MMVD patients. Representative criteria include: 1) **Mitral Regurgitation (MR) fraction**, which measures the degree of blood regurgitation from the left ventricle to the left atrium using color Doppler (Figure 3-8), 2) **LA/Ao ratio**, which evaluates the degree of left atrial enlargement by measuring the ratio of the left atrium to the aorta (Figure 3-10A), 3) **E-peak velocity (ventricular diastolic filling wave)**, which can measure the degree of blood inflow from the left atrium to the left ventricle using Doppler ultrasound at the mitral valve area (MVA), 4) **Mitral Regurgitation (MR) velocity**, which measures the speed of blood regurgitation from the left ventricle to the left atrium (Figure 3-10B), and 5) **left atrial (LA) diameter** measurement method. However, the fourth method is variable depending on the patient's condition or excitement, and the fifth method is not reliable for evaluating the degree of left atrial enlargement due to a large error range in the left atrial size criteria. More detailed information on these echocardiographic diagnostic evaluation methods for MMVD patients can be found in the Cardiac Ultrasound Diagnostic Guide section.

In human medicine, the prognosis is determined by measuring the thickness of the valve leaflets, but such criteria have not been established in animals. However, measuring the thickness of these leaflets is said to be useful in animals as well. The disadvantage of this method is that the leaflets appear thicker during ventricular diastole and the normal range varies between individuals. To accurately measure the thickness, it must be comparatively measured with the aortic valve to differentiate leaflet thickening. Also, **leaflet thickening on 2D ultrasound (Figure 3-7B; Video 3-1)** or **mitral valve prolapse (Figure 3-7A; Video 3-7)** where the **hinge point of the two mitral valve leaflets** protrudes into the left atrium are important findings of MMVD. As the valve thickens, **EPSS (E-point to septal separation)** is almost obliterated, and a flattening of the F-slope due to delayed valve opening and

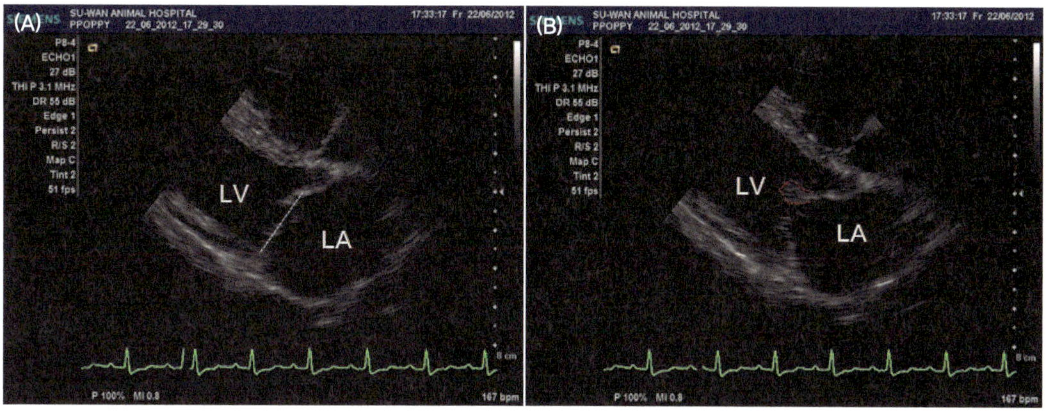

Figure 3-7 ▸ **2D Echocardiographic Findings of myxomatous mitral valve disease (MMVD).** (A) Mitral valve prolapse into the left atrium, (B) Thickening of the valve leaflets. LV: Left ventricle, LA: Left atrium.

closing can also be observed on M-mode (Figure 3-9B). Although MMVD is a systolic heart disease, %FS or %LVEF may appear normal or even increased due to the heart's compensatory response (Figure 3-9A; Video 3-4). Additionally, left atrial enlargement and regurgitant blood flow due to valve insufficiency can be confirmed (Figure 3-8; Video 3-2). Typically, the definitive diagnosis of MMVD can be made by diagnosing regurgitant blood flow from the mitral annulus into the left atrium using Doppler ultrasound (Figure 3-10B; Video 3-6). Furthermore, if the **LA/Ao ratio** (left atrial to aortic diameter ratio) in the aortic short axis view is ≥ **1.4:1**, it is suspected, and if it is ≥ **1.8:1**, it is confirmed (Figure 3-10A; Video 3-5).

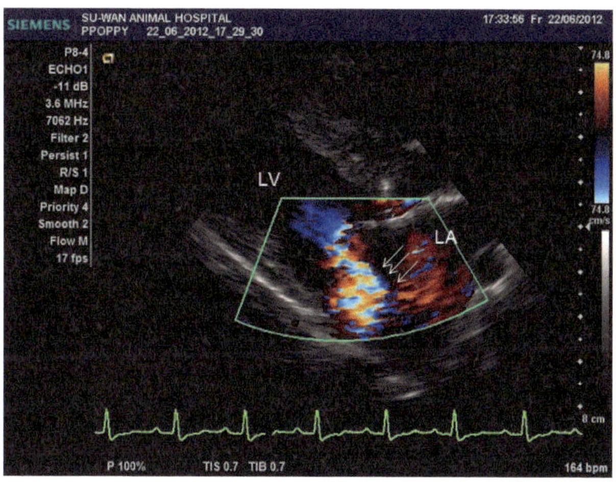

Figure 3-8 ▸ **Color Doppler Echocardiographic Findings of myxomatous mitral valve disease (MMVD).** Severe left atrial enlargement due to valve insufficiency and mosaic-patterned blood turbulence (arrow) regurgitating into the left atrium during systole due to mitral valve insufficiency. LV: Left ventricle, LA: Left atrium.

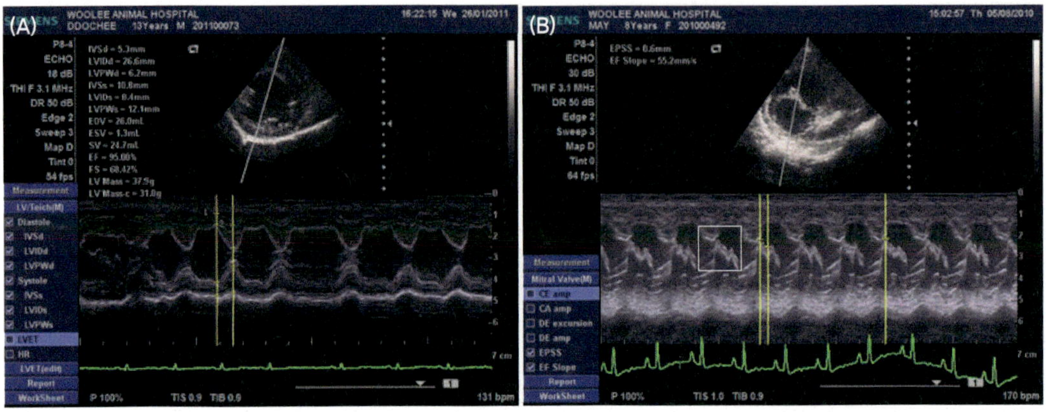

Figure 3-9 ▸ M-mode Echocardiographic Findings of myxomatous mitral valve disease (MMVD). (A) Left ventricular papillary muscle view: Although MMVD is a systolic heart disease, %FS and %LVEF actually show an increase due to the heart's compensatory response. (B) Left ventricular mitral valve view: As the valve thickens, EPSS is almost completely lost, and the F-slope becomes flattened due to delayed valve opening and closing.

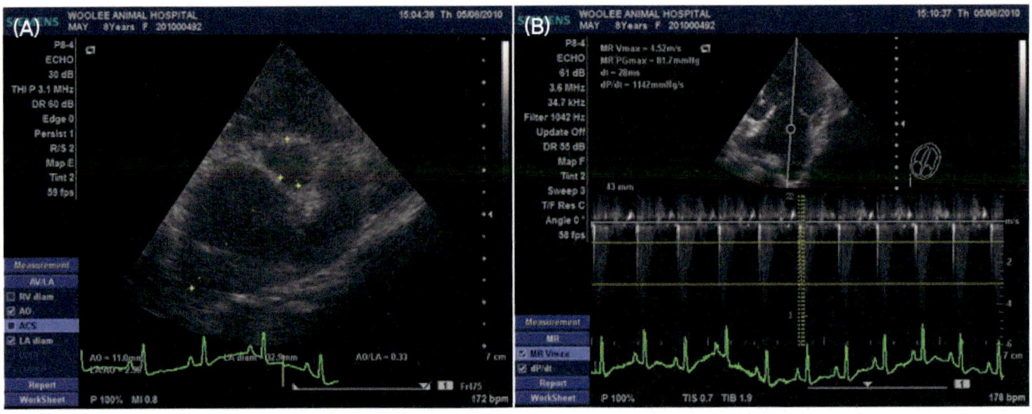

Figure 3-10 ▸ 2D (A) and Continuous Wave Doppler (B) Echocardiographic Findings of myxomatous mitral valve disease (MMVD). (A) The LA:Ao ratio is closely correlated with the progression of MMVD. If the left atrial to aortic diameter ratio (LA:Ao ratio) is 1.4:1 or higher, it suggests suspicion; if it's 1.8:1 or higher, it confirms the diagnosis. (B) MMVD can be confirmed by diagnosing the regurgitant flow from the mitral annulus into the left atrium using Doppler ultrasound.

Video Material 3-1 ▸ Myxomatous mitral valve disease. Right parasternal long axis 4 chamber view. Findings of thickened and prolapsed mitral valve leaflets.

Video Material 3-2 ▸ Myxomatous mitral valve disease. Right parasternal long axis 4 chamber view. Mosaic-patterned blood flow regurgitating into the left atrium (MR) due to mitral valve insufficiency.

Video Material 3-3 ▸ Myxomatous mitral valve disease (Same patient as Figure 10-2). Right parasternal long axis 4 chamber view. The patient also has tricuspid valve insufficiency, and mosaic-patterned blood flow regurgitating into the right atrium (MR) is also confirmed.

Video Material 3-4 ▸ Myxomatous mitral valve disease. M-mode echocardiography. Right parasternal short axis at the left ventricular papillary muscle level. The patient shows hyperdynamic movement of the septum to compensate for decreased cardiac contractility due to left ventricular dilation caused by mitral valve insufficiency.

Video Material 3-5 ▸ Myxomatous mitral valve disease. Right parasternal short axis 4 chamber view. Aortic valve level. LA/Ao ratio is being measured to confirm left atrial enlargement.

Video Material 3-6 ▸ Myxomatous mitral valve disease. Continuous wave Doppler. Left apical 4 chamber view. Measuring the regurgitant flow (MR) into the left atrium due to mitral valve insufficiency.

Video Material 3-7 ▸ Myxomatous mitral valve disease. Color Doppler ultrasound. Left apical 4 chamber view. Mosaic-patterned blood flow regurgitating into the left atrium (MR) formed due to mitral valve insufficiency.

3.5 Treatment and Prognosis

The treatment of myxomatous mitral valve disease remains a subject of ongoing debate. However, the main goals of treatment are to: 1) minimize cardiac workload, 2) correct the symptoms of congestive heart failure, 3) reduce regurgitant volume to minimize cardiac remodeling and optimize cardiac output, and 4) suppress and correct the occurrence of arrhythmias.

In human medicine, surgical procedures such as valvuloplasty (valve repair) or valve replacement therapy are performed. While mitral valve plastic surgery is performed for MMVD patients in some international centers, it has not yet been attempted in Korea. Additionally, transcatheter edge-to-edge mitral valve repair (TEER) using devices like the Canine Mitral V-Clamp is currently under clinical feasibility investigation in dogs with severe degenerative MR. These procedures and surgical treatments are often difficult to apply in dogs due to economic and technical limitations.

In contrast, atrial decompression, a procedure that creates an opening in the interatrial septum to shunt pressure overload from the left atrium to the right atrium, is gaining interest as a new treatment for patients with recurrent pulmonary edema. Currently, this procedure is performed at my hospital and by Dr. Allen at VCA Animal Hospitals in the US. Patients who have progressed to congestive heart failure due to MMVD should be treated according to the ACVIM diagnostic/treatment guidelines mentioned previously.

Some clinical and laboratory findings have been reported to be closely related to the prognosis of MMVD patients who have developed heart failure. Furthermore, the method of drug treatment is also known to be closely linked to patient survival rate and duration (e.g., pimobendan, ACEi). The dose of diuretics is also closely related to patient survival, with patients on high doses having a shorter life expectancy. Other factors closely related to the prognosis of MMVD patients include: the degree of exercise intolerance, the extent of cardiomegaly, the severity and extent of MR, the degree of left atrial enlargement (VHS score, left atrial size [LA/AO]), end-diastolic E/A-wave ratio, and decreased systolic function (increased end-systolic diameter). Recent studies also suggest that the vagus tone index (VTI) and the degree of tachycardia are closely related to MMVD prognosis. However, while the degree of syncope and the occurrence of arrhythmias increase the risk of sudden death, they are said to be less closely related to the progression of heart failure. Additionally, the occurrence of end-stage disease cachexia or increased kidney values due to cardiorenal syndrome are indicators of a poor prognosis. The breed is also closely related to the prognosis, as MMVD can be a benign change associated with aging in some breeds, while in others (especially Maltese, Shih Tzu), it can cause malignant heart failure.

The general survival period for MMVD patients is: 1) over 3 years for ISACHC I heart failure, 2) 1-3 years for ISACHC II heart failure, 3) up to 1 year for ISACHC IIIa, and 4) less than 6 months for ISACHC IIIb heart failure. The prognosis of this disease can vary greatly depending on the type of drug treatment, monitoring schedule, owner's commitment, and the patient's will to live.

3.6 Complications that Worsen the Progression of Heart Valve Disease

Cough due to Left Main Bronchial Compression: Severe LA enlargement can cause coughing by compressing the left main bronchus even without pulmonary congestion and edema. This can be confirmed on lateral radiographs. However, coughing due to other respiratory diseases (e.g., tracheal collapse) is also common in dogs with MMVD. The severity of coughing varies from a few intermittent coughs a day to continuous coughing throughout the day. Mild, intermittent coughing does not require drug treatment (as daily medication can be difficult and carries the risk of side effects). Furthermore, there is scant evidence that OTC cough medications help alleviate coughing symptoms in patients. Coughing caused by the left atrium compressing the main bronchus focuses on suppressing the cough reflex or alleviating the underlying cause of compression (left atrial enlargement). **Hydrocodone bitartrate (0.2-0.3 mg/kg, PO, q6-12h), butorphanol (0.55-1.1 mg/kg, PO, q6-12h), or dextromethorphan (0.5-2 mg/kg, PO, q6-8h)** may help alleviate coughing in some cases. Patients with concomitant tracheal collapse or small airway disease may benefit from bronchodilators or short-term steroid treatment. Various **xanthine derivatives** (e.g., **aminophylline [8-11 mg/kg, PO, q6-8h], theophylline [sustained-release 20 mg/kg, PO, q12h] or oxytriphylline [SR 25-30 mg/kg, PO, q12h]**) are commonly used as bronchodilators. However, the efficacy of these drugs varies greatly among individuals. **Beta-2 agonists** such as **terbutaline and albuterol** should be used with caution in MMVD dogs (as these drugs can stimulate myocardial beta-2 receptors, increasing heart rate and contractility).

Theoretically, coughing due to left main bronchial compression can be alleviated by reducing LA and LV size. Therefore, reducing MR or pulmonary venous pressure can decrease LA size and thereby alleviate coughing symptoms. Reducing afterload can decrease the amount of MR, and for this purpose, arterial vasodilators like **amlodipine and hydralazine** can be used, and **ACEi**, though less effective, can also be used. Furthermore, the inodilator **pimobendan** is known to be effective in reducing heart size in MMVD patients. Diuretic monotherapy can also reduce LV size by decreasing blood volume, thereby alleviating MR. However, diuretics activate the RAAS and can cause electrolyte imbalances with long-term use. Therefore, diuretics should only be used to alleviate coughing symptoms that are not relieved by antitussives, steroids, or vasodilators.

Ruptured chordae tendineae (RCT) is a major factor that can suddenly cause or worsen pulmonary congestion and edema in all patients with valve disease. Such chordal rupture is very rare in cats. Chordal rupture typically occurs due to extreme stress or exercise (e.g., chasing cats or birds). The most serious complications arise when the chordal tendon attached to the A1 septal leaflet (anterior leaflet) ruptures. In this case, **volume overload** rapidly occurs in the atrium and ventricle, leading to acute pulmonary edema and patient death. However, if chordal tendons supporting other septal leaflets (A2, A3) or the free-wall leaflet (posterior leaflet) rupture, clinical symptoms may be minimal or absent. Severe chordal rupture leads to a sudden exacerbation of mitral regurgitation (MR). When MR suddenly worsens, left atrial and pulmonary venous pressures rapidly increase, causing **acute pulmonary edema and pulmonary hypertension**. However, in MMVD without chordal rupture, MR worsens slowly, so clinical symptoms do not deteriorate rapidly. During a physical examination of a patient with sudden chordal rupture, a much lower grade murmur than in chronic MR patients or an S3 gallop may be heard, and jugular distention or pulsation may appear. Diagnostic imaging findings vary depending on the progression of valve disease before the chordal rupture. Severe MR can be confirmed on Doppler echocardiography, and **frail mitral valve** movement, independent of the cardiac cycle, can be observed. These patients require emergency intensive care to stabilize symptoms, and continuous drug therapy for heart failure is necessary even after rescue. Unless it is an A1 septal leaflet rupture, most chordal ruptures can be treated with appropriate drug management.

Right Heart Failure due to Pulmonary Hypertension: In many patients who have suffered from valve disease for a long time, right heart failure also occurs. This occurs when tricuspid insufficiency is concurrent (Video 3-3), pulmonary hypertension worsens, or both. In MMVD patients, pulmonary hypertension occurs due to persistent increases in left atrial pressure and pulmonary venous pressure, and co-occurring chronic bronchial disease also exacerbates pulmonary hypertension. Because the right ventricle is thinner than the left ventricle, it can relatively accommodate a large volume of blood (volume overload) but cannot tolerate pressure overload. Therefore, even a slight increase in pulmonary artery pressure can significantly decrease the right ventricular stroke volume (SV). Patients with pulmonary hypertension easily tire with even a slight increase in activity and may even faint in severe cases. Clinical symptoms of right heart failure such as **ascites, pleural effusion, and hepatosplenomegaly** appear on physical examination. The presence and severity of pulmonary hypertension can be indirectly measured using Doppler echocardiography. **Sildenafil (1-4 mg/kg q8-12h PO)** or **tadalafil (0.5-2 mg/kg q24h PO)** can be used for the treatment of pulmonary hypertension. A less expensive alternative to sildenafil is **pimobendan**. This drug has experimentally shown an effect of lowering pulmonary artery pressure in dogs with induced PHT (by inhibiting PDE III). In drug crossover studies conducted on MMVD patients, pimobendan also had an effect of alleviating PHT. Therefore, MMVD patients with chronic CHF can benefit from these drug treatments. Bronchodilators may be indicated. **Methylxanthine and beta-2 agonists** can be used. However, beta-2 agonists should be used with caution in MMVD patients. Aggressive diuretic treatment may be necessary to control ascites. Fluid removal through abdominocentesis provides only temporary relief, so additional drug prescriptions can slow the rate of fluid accumulation.

Left atrial enlargement due to valve disease makes the left atrial wall thinner and more vulnerable to pressure overload (e.g., chordal rupture or trauma). **Endocardial splitting** is frequently observed in MMVD patients with long-term survival. These lesions can suddenly cause left atrial rupture, leading to **hemopericardium, cardiac tamponade, and sudden death**. In individuals who survive such events, acute ascites, collapse, or marked exercise intolerance may occur due to symptoms of right heart failure caused by cardiac tamponade. Refer to textbooks or review materials for detailed clinical symptoms and diagnostic methods for pericardial effusion. The prognosis of hemopericardium due to atrial rupture is very poor, but immediate pericardiocentesis can minimize cardiac dysfunction. However, pericardiocentesis sometimes worsens bleeding from atrial rupture. As a last resort, surgical repair of the atrium through thoracotomy can be considered, but the success rate is not high.

3.7 Systemic Complications Due to Heart Valve Disease

Cardiorenal Syndrome (CRS)

Cardiorenal Syndrome (CRS) is a condition where renal dysfunction occurs secondarily to heart failure, characterized by diuretic resistance, co-occurrence of heart failure and renal failure, or worsening of renal function due to aggressive heart failure treatment. When both renal failure and heart failure occur simultaneously, heart failure exacerbates renal function, and deteriorated renal function in turn adversely affects heart failure, creating a vicious cycle. Although the definition and classification are not yet clearly established in veterinary medicine, the classification in human medicine is as follows: Type 1-Acute cardiorenal syndrome (acute kidney injury caused by acute heart failure), Type 2-Chronic cardiorenal syndrome (gradual kidney damage leading to renal failure caused by chronic heart failure), Type 3-Acute renocardiac syndrome (acute heart failure caused by acute kidney injury), Type 4-Chronic renocardiac syndrome (chronic heart failure caused by chronic kidney disease), Type 5-Secondary cardiorenal syndrome (cardiac and renal dysfunction occurring secondarily due to diseases such as sepsis, amyloidosis, or diabetes).

Pathogenesis: A retrospective study in human medicine reported that 60% of 755 patients with chronic heart failure also had chronic kidney disease, and a close relationship between glomerular filtration rate and the prognosis of chronic heart failure has been reported. Recent veterinary studies also report that azotemia and kidney damage occur more frequently with increasing severity of heart failure. Since the kidneys receive 25% of the blood pumped from the heart, these two organs share various homeostatic mechanisms hemodynamically and neuro-hormonally. The problem is that while the homeostatic responses of the two organs do not affect each other in normal conditions, these homeostatic responses adversely affect the other organ in the presence of renal failure or heart failure (Figure 3-11). For example, in congestive heart failure, diuretic administration to reduce left atrial volume overload greatly helps alleviate heart disease symptoms, but it reduces renal blood flow due to decreased circulating blood volume, thereby decreasing glomerular filtration rate and impairing renal function. If the patient has renal failure, decreased renal perfusion can further worsen renal failure. Conversely, patients with severe renal failure often have high intraglomerular pressure and glomerular hypertension due to glomerular fibrosis, etc. Ultimately, this glomerular hypertension often causes systemic hypertension. In a state of systemic hypertension, pressure overload on the left ventricle increases, increasing the workload on the left ventricular wall and increasing wall stress. If not properly treated, the left ventricular wall undergoes concentric hypertrophy (a compensatory response of the heart to hypertension characterized by significantly thickened left ventricular walls and an unexpanded ventricular lumen) and eventually leads to heart failure. However, in severe MR dogs, the increased systemic blood pressure may not cause concentric hypertrophy because the increased LV systolic pressure (due to increased arterial pressure) would be released through severe MR.

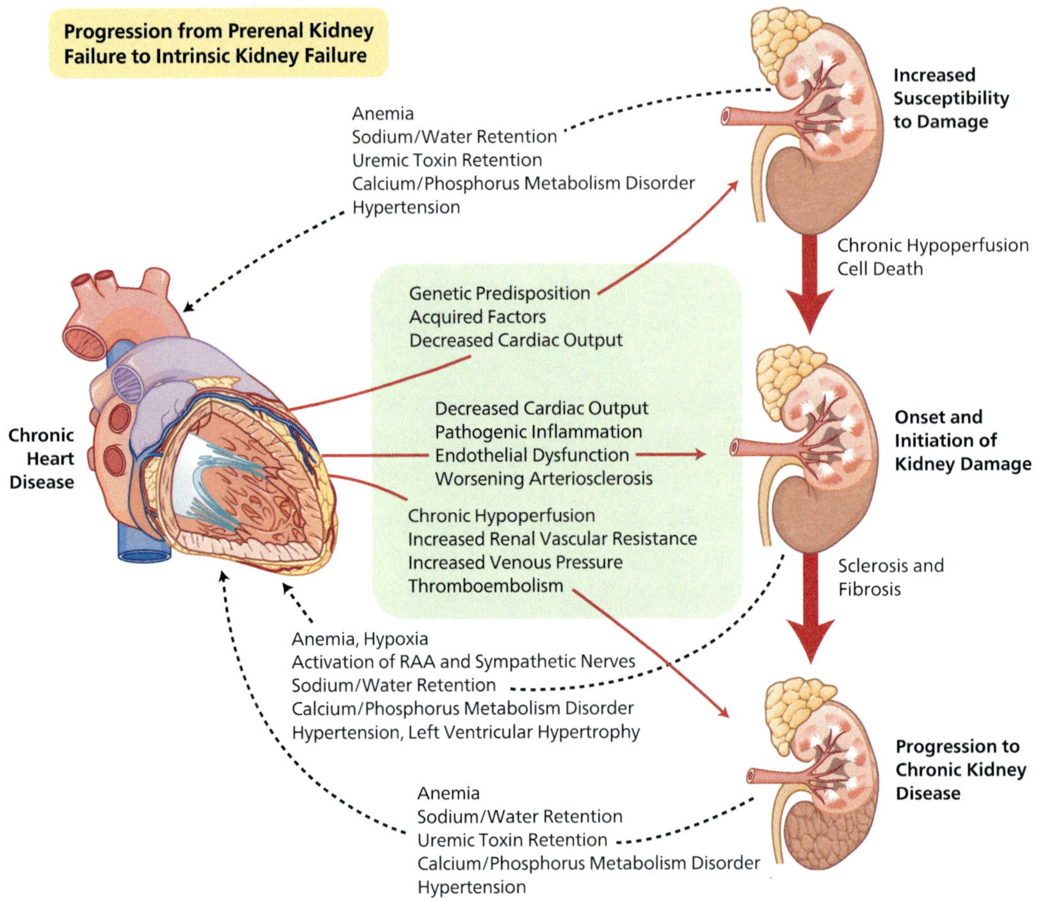

Figure 3-11 ▸ Pathogenesis of Cardiorenal Syndrome. Anemia, sodium and water retention, uremic toxin retention, calcium-phosphate imbalance, and hypertension caused by kidney disease worsen chronic heart disease (dashed line). Decreased cardiac output, hypoxic injury, hypoperfusion pressure, and thrombosis caused by heart disease worsen kidney disease (red line).

Long-term use of ACEi like enalapril in heart failure patients often leads to an increase in blood creatinine levels. This is because ACEi mainly dilate the efferent arteriole in the kidneys, reducing intraglomerular pressure and leading to renal hypoperfusion. This eventually reduces the glomerular filtration rate and can increase blood creatinine levels. However, prospective and retrospective studies on various ACEi in heart failure patients show that the positive effects of ACEi on cardiac and renal function are greater. ACEi slows the progression of renal parenchymal fibrosis and has a vasodilatory effect, reducing glomerular vessel damage and alleviating proteinuria. The negative effects of ACEi on the kidneys are also temporary, and a study examining the long-term efficacy of enalapril reported that the incidence of azotemia in heart failure patients administered a regular dose of enalapril (0.5mg/kg) was lower than in the control group.

Treatment and Prognosis of CRS: The treatment and management of cardiorenal syndrome patients are very challenging, and effective treatment methods are not yet well-established. Ultimately, regular monitoring of renal function is crucial under the assumption that all patients receiving long-term heart failure treatment will experience renal function decline.

The most effective treatment for cardiorenal syndrome patients is prevention. Therefore, establishing an appropriate heart failure treatment strategy tailored to the patient's heart failure progression and, above all, optimizing the diuretic dose is paramount. Strategies include: administering small, divided doses of diuretics rather than large doses at once (BID-TID), combining with or switching to a different diuretic if resistance to one diuretic is observed, and adding other heart failure medications to reduce diuretic use (e.g., spironolactone, nitrate patch, amlodipine). Most importantly, closely monitoring the patient's response to the diuretic dose and using the optimal diuretic dose for the patient is crucial.

As part of periodic kidney monitoring, measuring the renal resistive index (RI) via ultrasound can help assess whether renal blood flow resistance is increasing (RI $\geq$ 0.7 suggests renal failure; Figure 3-12). However, once chronic kidney disease has developed and the patient presents, it is difficult to apply standard renal failure treatment (aggressive fluid therapy). Aggressive fluid therapy in volume-overloaded patients can worsen congestive heart failure and cause pulmonary edema. Therefore, in such cases, lower doses of fluid therapy than for typical chronic kidney disease patients (administering only maintenance fluid, 30 mL/kg + 70 mL per hour at 5-10 mL/kg/hr) or performing peritoneal dialysis or hemodialysis may be effective. However, the most important thing is to educate the owner that periodic monitoring is essential because patients with heart failure are at high risk of developing kidney injury. Regular monitoring of blood tests (creatinine and SDMA), urine tests (urine specific gravity), UPC tests, and blood pressure is necessary to prevent the pa-

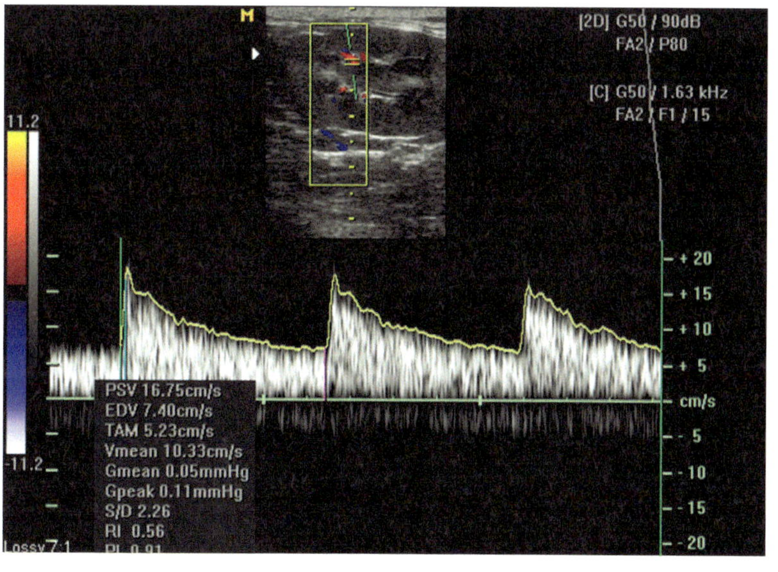

Figure 3-12 ▸ Resistive Index Examination Method for Measuring Pressure and Load in Renal Vessels.

tient from presenting in an azotemic state.

Generally, the prognosis varies greatly depending on the severity of heart failure. Most patients have previously received long-term diuretic prescriptions, so the long-term prognosis is generally not good.

Pancreatitis

Pancreatitis is the most common exocrine pancreatic disease in dogs. In MMVD patients, pancreatitis is one of the representative systemic complications. Poor perfusion and congestion due to heart failure seem to increase the incidence of pancreatitis. The brain, kidneys, and pancreas are typical organs susceptible to ischemic damage. As previously mentioned, just as the incidence of cardiorenal syndrome is high in chronic heart failure patients, the incidence of pancreatitis has also been proven to be high in dogs with heart failure. In 2015, my laboratory investigated serum canine pancreatic-specific lipase (cPL) levels in MMVD patients at different stages of heart failure. In this study, serum cPL levels were: 54.0 μg/L (IQR: 38.0–78.8 μg/L) for the normal control group, 55.0 μg/L (IQR: 38.3–88.8 μg/L) for the ISACHC I group, 115.0 μg/L (IQR: 45.0–179.0 μg/L) for the ISACHC II group, and 223.0 μg/L (IQR: 119.5–817.5 μg/L) for the ISACHC III group. cPL levels were higher with increasing stages of heart failure, and a close correlation was observed with LA/Ao ratio (r = 0.597; P = .000) and heart failure severity (r = 0.530; P = .000) (Figure 3-13).

This study also found that the number of individuals with borderline cPL levels (201–399 μg/L) suggesting pancreatitis and those with levels (≥400 μg/L) indicating pancreatitis significantly increased in the symptomatic ISACHC II and III groups. Interestingly, very few individuals showed clear symptoms like acute pancreatitis, with most exhibiting mild/intermittent gastrointestinal symptoms. Recent clinical experience indicates a high co-oc-

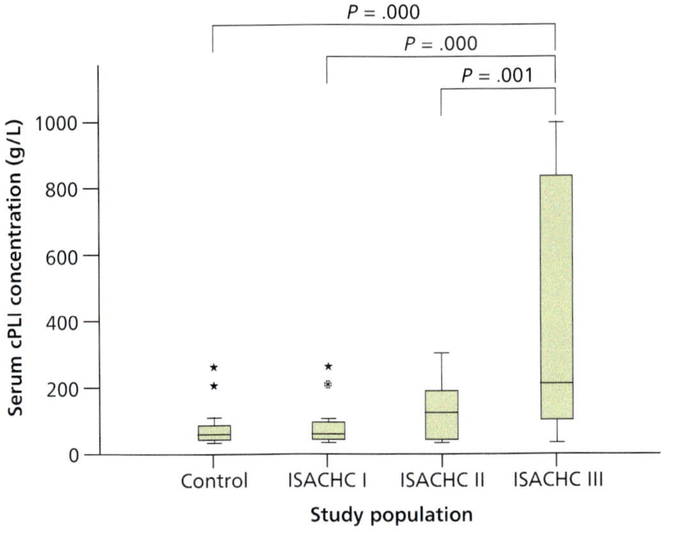

Figure 3-13 ▸ Serum cPL Concentration Distribution in MMVD Patients.

currence of CRS and pancreatitis in heart failure patients. These patients often showed clear pancreatitis symptoms and renal failure, making management challenging. Therefore, early detection and appropriate treatment before clinical pancreatitis develops are crucial. In my practice, I administer pancreatic enzyme supplements to individuals who test positive for cPL (via SNAP or Spec cPL tests). For most patients, levels normalize within 1-4 months with enzyme supplementation alone. If levels remain elevated, additional dietary changes are needed. Since most renal prescription diets are high in fat, feeding them to these patients is not advisable.

3.8 Echocardiographic Diagnostic Guidelines for Mitral Valve Insufficiency Patients

Accurate diagnosis of **mitral valve insufficiency (MVI)** requires strict imaging diagnostic guidelines. In addition to imaging data, careful evaluation of the patient's **medical history, physical examination, electrocardiogram, and laboratory findings** is crucial for precise diagnosis and assessment of the patient's condition. Clinically, **echocardiography** is a very important examination for understanding valve lesion morphology, degree of mitral regurgitation, extent of cardiac remodeling, presence of diastolic and systolic ventricular dysfunction, and evidence of elevated blood pressure in the diagnosis of MVI. Therefore, when performing echocardiography, evaluation of valve degeneration and cardiac remodeling, pulmonary artery pressure, left ventricular function during diastole and systole, and increased ventricular filling pressure must be performed.

Assessment of MR Severity

Evaluation of **mitral regurgitation (MR)** flow volume using **color Doppler and continuous wave Doppler** is the standard diagnostic method for MMVD, allowing for semi-quantitative/quantitative assessment of disease severity in dogs.

Semi-quantitative Assessment of MR

In humans, the severity of MR can be quantitatively assessed by measuring **regurgitant volume (R Vol)** or **effective regurgitant orifice area (EROA)**. However, in dogs, due to the varied direction and shape of the regurgitant jet and the difficulty in precisely identifying the EROA location with ultrasound (which is also time-consuming), these methods are rarely applied in clinical practice. For this reason, **semi-quantitative assessment methods** (e.g., **the maximal ratio of the regurgitant jet area signal to LA area [ARJ/LAA ratio] or vena contracta** [VC]) are primarily used in veterinary medicine to assess MR severity.

The **ARJ/LAA ratio** can be easily measured with **color Doppler imaging (CDI)** and has high reproducibility in dogs. In canine MMVD, MR severity is classified as mild (<30%), moderate (30-70%), and severe (>70%) based on the ARJ/LAA ratio (Figure 3-14A). Unfortunately, the ARJ/LAA ratio is influenced by systolic blood pressure, left atrial pressure, or the spatial arrangement of the MR jet. It is also affected by echocardiography machine

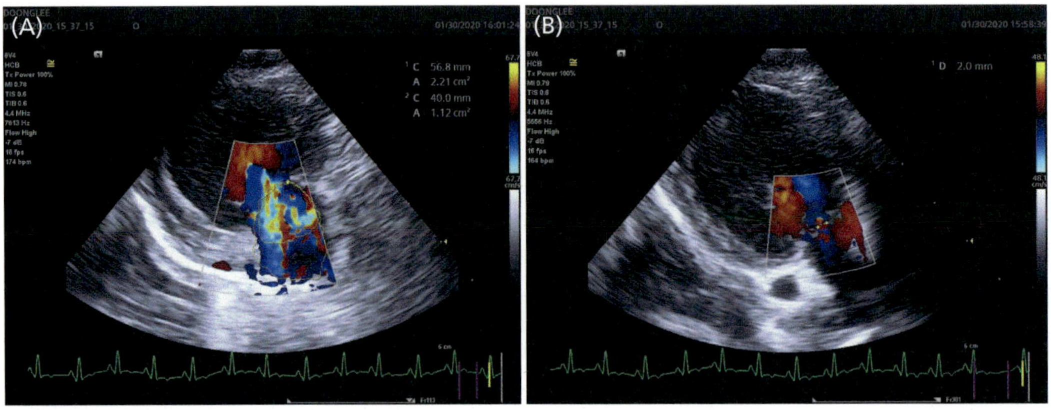

Figure 3-14 ▸ (A) Semi-quantitative measurement of MR: maximal ratio of the regurgitant jet area signal to left atrial area (ARJ/LAA ratio). In this patient, ARJ (1.12 cm^2) / LAA (2.12 cm^2) = 53%, indicating moderate MR. (B) Vena contracta (VC) refers to the narrowest part of the mitral orifice where the MR jet forms; in this patient, it is 2 mm.

settings (e.g., pulse repetition frequency and gain settings). Furthermore, several studies have shown that the ARJ/LAA ratio is not closely correlated with the severity of MMVD (especially in ACVIM B2 and C groups).

Vena contracta (VC) refers to the narrowest part of the mitral orifice where the MR jet forms and can be measured with **CDI**, allowing for assessment of MR severity (Figure 3-14B). Although accurate measurement takes some time, it is technically easy to perform. However, this method is prone to measurement errors in cases with **dynamic regurgitant orifices**. Nevertheless, several studies have reported a close correlation between VC and VC-related indices with MR severity and survival time. One study in MMVD dogs found that the ratio of aortic diameter to VC size (right parasternal measurement VCR/Ao and left apical measurement VCL:Ao) was 0.21 ± 0.14 and 0.24 ± 0.12, respectively, with **VCL:Ao > 0.24** being closely associated with survival time (hazard ratio 4.87).

Quantitative Assessment of MR

The **flow convergence area measurement method using proximal isovelocity surface area (PISA)** is the standard method for quantifying MR in humans. Although one study suggested that the PISA method is reproducible in dogs, it is a time-consuming method that requires optimal image acquisition. The **regurgitation fraction (RF)** obtained by the PISA method can be used to classify MR severity in MMVD patients as mild (<30%), moderate (30-75%), and severe (>75% of RF). Several studies have reported a close correlation between RF and MMVD severity and other echocardiographic parameters (e.g., LA/Ao, systolic PA pressure). However, one study showed that MR quantification by the PISA method showed various severities even in asymptomatic groups (approximately 33% of asymptomatic groups showed moderate to severe RF values). For these reasons, this method is not widely used in clinical practice.

Measurement of Left Atrial and Left Ventricular Enlargement

Chronic and hemodynamically significant MR can cause volume overload in the left atrium and left ventricle, leading to their enlargement. The extent of this left atrial and left ventricular enlargement can be measured using **LA to aortic root ratio (LA/Ao), indexed LA diameter (iLA), LV end-diastolic internal dimension to aortic root ratio (LVIDd/Ao), and normalized LV end-diastolic internal dimension** (LVIDdN).

Assessment of Left Atrial Enlargement

Left atrial diameter measurement is one of the most useful indicators for assessing the prognosis of MMVD patients. These indicators serve as key criteria for initiating drug prescriptions in asymptomatic patients. They are also excellent echocardiographic indicators that can predict the occurrence of left heart failure. Chronic and hemodynamically significant MR causes volume overload in the left atrium, leading to its enlargement. Since MR expands both the left atrium and left ventricle, measuring the diameters of the left atrium and left ventricle is useful for predicting the progression stage of MMVD and the risk of de-

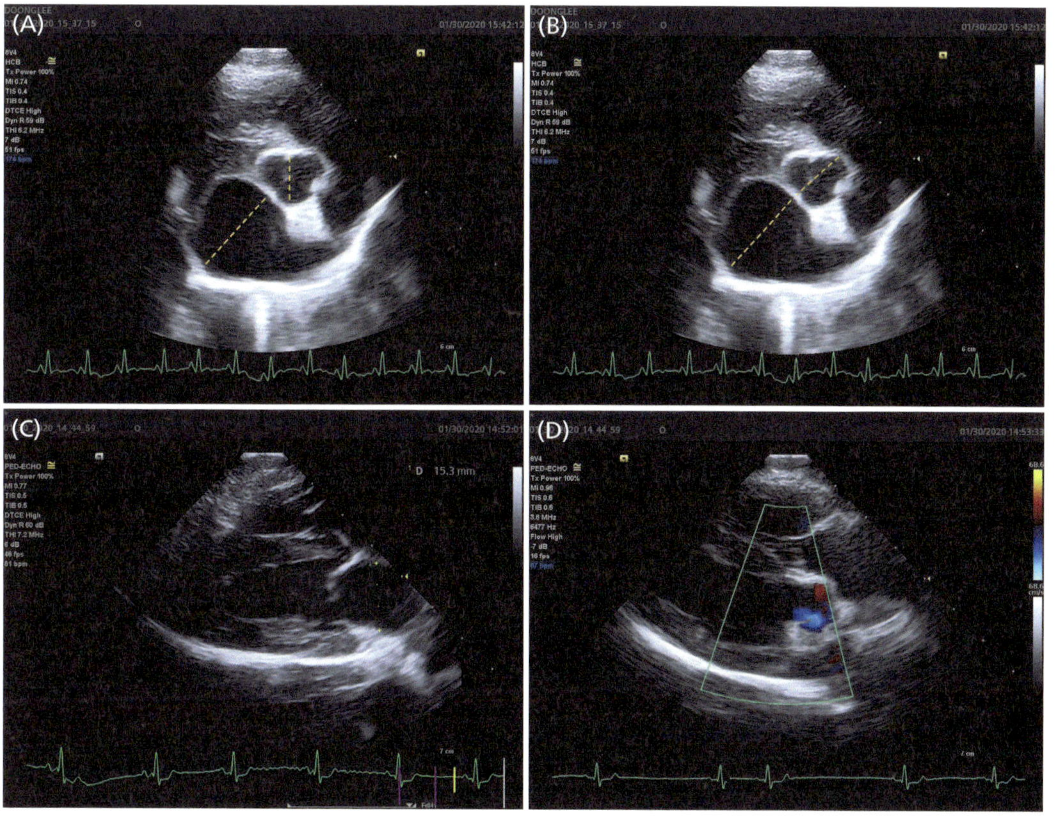

Figure 3-15 ▸ Various methods for measuring left atrial to aortic root diameter ratio (LA/Ao). (A) Method 1 for measuring from the aortic short-axis view (LA/AoSx) (see text). (B) Method 2 for measuring from the aortic short-axis view (LA/AoSx) (see text). (C and D) Method for measuring from the right parasternal long-axis view (LA/AoLx) (see text).

veloping congestive heart failure. In one study on MMVD patients, when LA, LV, and aortic root were measured from the right parasternal short-axis view (RPLx), the average values of **LA/LV, LV/Ao, and LA/Ao were 1.1±0.09, 2.7±0.5, and 3.0±0.5**, respectively, showing significant differences compared to the normal control group's values of 1.0±0.05, 2.1±0.2, and 2.1±0.1. In this study on MMVD patients, LA/LV, LV/Ao, and LA/Ao showed high reproducibility and applicability in evaluating left atrial and left ventricular enlargement.

The standard echocardiographic method for assessing the degree of left atrial enlargement is to measure **LA/Ao** in the aortic short-axis view (Figure 3-15A and B). In patients with MMVD, the extent of left atrial enlargement was closely correlated with **congestive heart failure (CHF)** and survival time. In one study of 558 MMVD patients, among all echocardiographic parameters, only **LA/Ao >1.7** was closely associated with prognosis. Although LA/Ao is the most important indicator for determining the extent of left atrial enlargement, the measurement method itself is not standardized. The first commonly used method in clinical practice is as follows: Step 1 - Measure the aortic annulus diameter from inner edge to inner edge (usually the right coronary cusp to the commissures of the left and noncoronary cusps). Step 2 - Measure the maximum left atrial diameter (LAmax) from the trailing edge of the aortic annulus to the distal left atrial free wall in early diastole (Figure 3-15A). However, many other studies measure LA/Ao using a different method (measure the inner diameter between the noncoronary aortic valve cusps and the right coronary aortic valve cusps when the aortic valve is closed in the aortic short-axis view, and then measure the left atrial diameter using the method mentioned above; Figure 3-15B). My lab's research showed that the aortic diameter measured by the first method for LA/Ao was longer than that measured by the latter method. Therefore, the LA/Ao ratio measured by the first method was smaller than that measured by the latter method. Since LA/Ao is used as the most important echocardiographic parameter for patient classification in many drug trials and clinical studies, a unified method for measuring aortic diameter is needed. Although **LA/Ao in the short-axis view (LA/AoSx)** is primarily measured, LA/Ao can also be measured in the long-axis view (LA/AoLx). These measurement methods minimize errors due to differences in aortic or left atrial measurement methods because they use standardized examination procedures (Figure 3-15C, D). Several studies have shown that **LA/AoLx >2.6 and LA/AoSx >1.6** are criteria for defining hemodynamically significant MMVD and are indicators for initiating treatment drugs (e.g., **enalapril and pimobendan**).

The **indexed LA diameter (iLA)** is a good indicator for assessing left atrial enlargement. iLA can be calculated using the following formula: LA diameter (mm)/(0.795 × body weight [kg])1/3. One study found that **iLA diameter** had high sensitivity and specificity in differentiating healthy dogs from asymptomatic heart failure patients, but low specificity in differentiating clinical from asymptomatic heart failure. A cutoff value of **iLA >12.7** was a useful indicator for predicting the risk of heart failure in MMVD patients.

Since the left atrium has a true **3-dimensional (3D) structure**, left atrial enlargement can occur in any direction (e.g., craniocaudal, mediolateral, dorsoventral). Therefore, LA/Ao measured in 2D images may be overestimated or underestimated depending on the direction of left atrial enlargement. For this reason, linear methods (e.g., LA/AoSx, LA/AoLx, iLA) may not provide accurate and consistent measurements of left atrial diameter. One study showed that 3D echocardiography for left atrial volume provided more accurate measurements for detecting mild left atrial enlargement compared to 2D measurement methods. Furthermore, human studies have shown that 2D measurements of left atrial diameter and area have a weaker correlation with 3D measurements. However, these measurement methods are more time-consuming and require substantial off-line analysis. Additionally, human studies have reported that 3D measured left atrial volumes are underestimated compared to MRI measured left atrial volumes.

Area-length method (ALM; Figure 3-16A) or **monoplane/biplane Simpson's modified method of discs (SMOD; Figure 3-16B)** are used for measuring left atrial area for left atrial volume measurement. In humans, biplane measurement using 4-chamber and 2-chamber views is primarily used to measure left atrial volume. One study compared 3D-LA volume measurement with two 2D LA volume measurement methods (SMOD and ALM). SMOD and ALM methods were found to underestimate or overestimate left atrial volume by 7% and 24% respectively compared to the 3D method (Figure 3-16C). A human study found that the SMOD method measured left atrial volume smaller than the 3D method. Therefore, if 3D measurement is not available, the **2D-SMOD** method is more useful for left atrial volume measurement. One study used the ALM method to calculate the echocardiographic parameters of **left atrial maximal volume (LAVmax)** and **left atrial minimal volume (LAVmin)**. These indicators were closely related to patient survival when only cardiac-related mortality was considered. In one study, **LAVmax >3.53 mL/kg** was closely associated with high mortality in MMVD patients.

Indicators for evaluating left atrial function were useful in assessing disease severity in MMVD patients. The left atrium performs three functions: 1) preserving pulmonary venous return during ventricular systole (atrial diastole), 2) acting as a conduit to send blood from the left atrium to the left ventricle in early ventricular diastole, and 3) acting as an active pump to send 15-30% of blood to the ventricle in late ventricular diastole (atrial systole). One study set reference ranges for left atrial functional indicators using 2D area measurements and investigated whether they could differentiate ACVIM B asymptomatic group from ACVIM C symptomatic group. This study evaluated the following four left atrial functional indicators: **LA expansion index for LA reservoir, LA active emptying fraction for LA contractile, LA passive emptying fraction for LA conduit, and LA total emptying fraction for LA reservoir**. In this study, all left atrial functional indicators except LA passive emptying worsened with the progression of heart failure. However, these indicators alone could not differentiate asymptomatic from symptomatic groups.

Assessment of Left Ventricular Enlargement

In patients with chronic and hemodynamically significant MR, left ventricular diameter is often increased. The degree of left ventricular enlargement is closely related to the severity of the disease in MMVD patients. Left ventricular diameter can be measured in **end-systole and end-diastole** using **2D or M-mode echocardiography** (Figure 3-18). As mentioned above, the **LV internal dimension** was significantly wider in MMVD patients. However, the

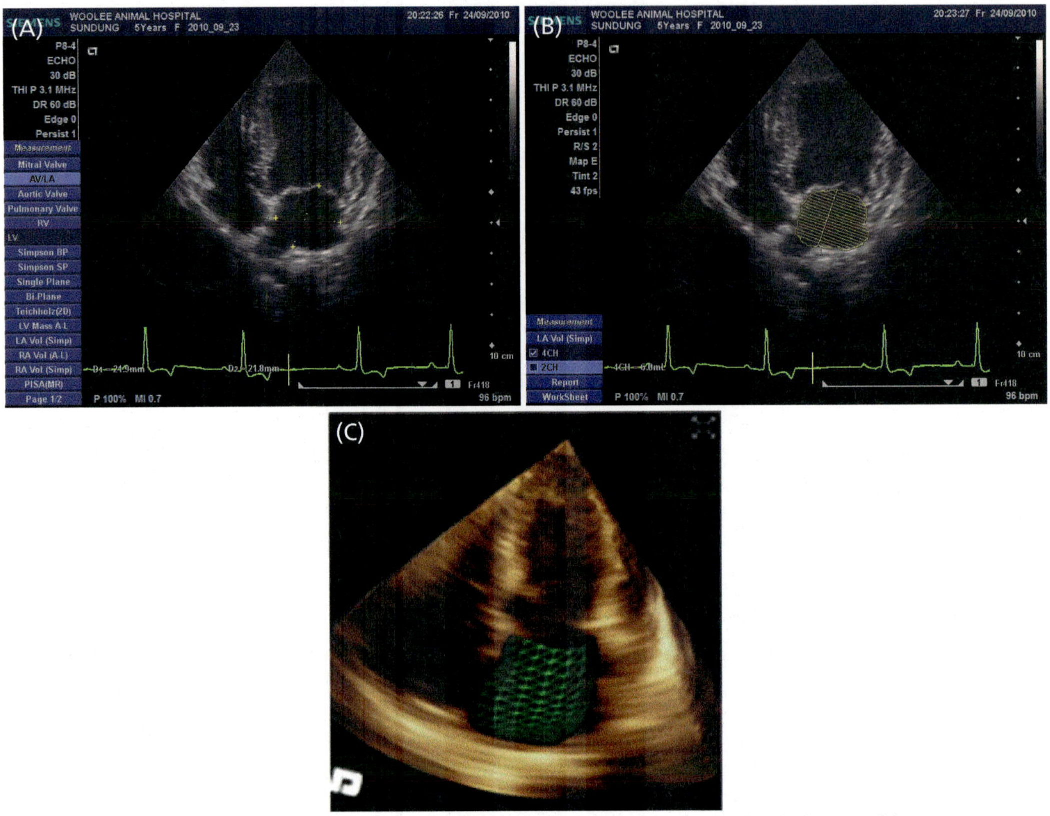

Figure 3-16 ▸ Three methods for measuring left atrial volume from the left apical long-axis view. (A) Area length method (ALM), (B) Simpson's modified method of discs (SMOD), (C) Real-time 3-dimensional echocardiography (RT3DE).

LV internal dimension varied greatly by breed. A **bodyweight-indexed (normalized) LV internal dimension** measurement method was developed, considering the correlation between M-mode measurements and body weight in healthy adult dogs. These values can be calculated as follows:

- **LV end diastolic diameter normalized for body weight (LVIDdN) = LVIDd (cm)/ weight (kg)$^{0.294}$**
- **LV end systolic diameter normalized for body weight (LVIDsN) = LVIDs (cm)/ weight (kg)$^{0.315}$**

LVIDdN ≥1.7 indicates heart enlargement of ACVIM B2 or higher, requiring cardiac medication. Several studies have shown that **end-diastolic volume index (EDVI)** and **systolic volume index (ESVI)**, which are left ventricular diameters corrected for body surface area (BSA), are useful in assessing the degree of systolic dysfunction in MMVD patients. EDVI and ESVI can be measured using the **Teichholz formula**:

- **EDVI = $[7\times(\text{LVIDd})^3/(2.4+\text{LVIDd})]/\text{BSA}$**
- **ESVI = $[7\times(\text{LVIDs})^3/(2.4+\text{LVIDs})]/\text{BSA}$**

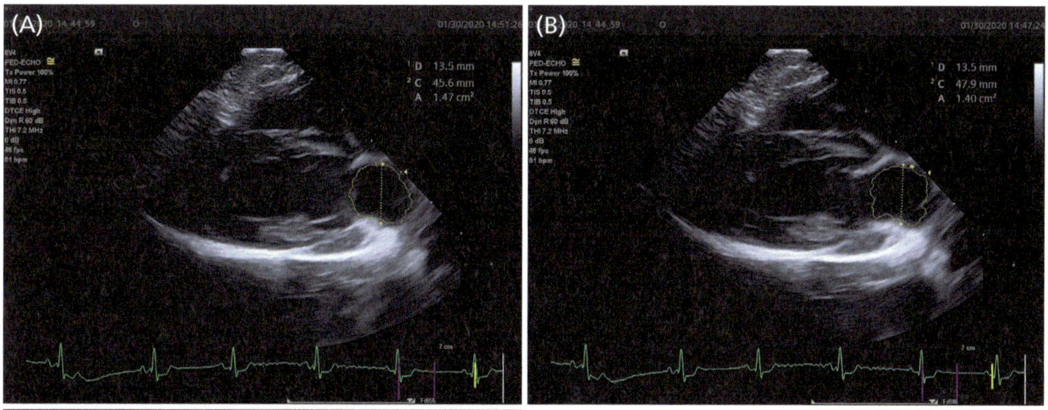

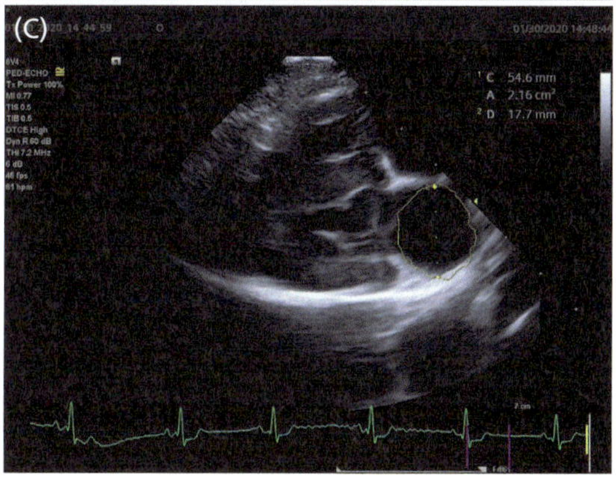

(D)

	Function assessed	Index	Abbreviation	Formula
Linear	LA reservoir	LA expansion index	LAEXP	(LAMAX-LAMIN)/LAMIN×100
	LA contractile	LA active emptying fraction	LAACT	(LAP-LAMIN)/LAP×100
	LA conduit	LA passive emptying fraction	LAPAS	(LAMAX-LAP)/LAMAX×100
	LA reservoir	LA total emptying fraction	LAFT	(LAMAX-LAMIN)/LAMAX×100
Area	LA reservoir	LA expansion index	LAAEXP	(LAAMAX-LAAMIN)/LAAMIN×100
	LA contractile	LA active emptying fraction	LAAACT	(LAAP-LAAMIN)/LAAP×100
	LA conduit	LA passive emptying fraction	LAAPAS	(LAAMAX-LAAP)/LAAMAX×100
	LA reservoir	LA total emptying fraction	LAAFT	(LAAMAX-LAAMIN)/LAAMAX×100

LA, left atrium; LA_{max}, maximal left atrial dimension; LA_{min}, minimal left atrial dimension; LA_{P}, left atrial dimension at the onset of the p-wave on the ECG; LA_{EXP}, left atrial expansion index (based on linear measurements); LA_{ACT}, left atrial active emptying fraction (based on linear measurements); LA_{PAS}, left atrial passive emptying fraction (based on linear measurements); LA_{FT}, left total atrial emptying fraction (based on linear measurements); LAA_{max}, maximal left atrial area; LAA_{min}, minimal left atrial area; LAA_{P}, left atrial area at the onset of the p-wave on the ECG; LAA_{EXP}, left atrial expansion index (based on area measurements); LAA_{ACT}, left atrial active emptying fraction (based on area measurements); LAA_{PAS}, left atrial passive emptying fraction (based on area measurements); LAA_{FT}, left atrial total emptying fraction (based on area measurements).

Figure 3-17 ▶ Methods for evaluating left atrial function on 2D echocardiography. (A) Minimal left atrial diameter/area (late ventricular diastole), (B) Maximal left atrial diameter (early ventricular diastole), (C) Mid left atrial diameter/area (diastasis), (D) Formula for calculating left atrial functional index.

EDVI and ESVI can be calculated using BSA. Normal ranges for EDVI and ESVI are <100 mL/m^2 and <30 mL/m^2, respectively. One study evaluated the accuracy of ESVI using two measurement methods (e.g., the geometric [GM, based on Teichholz formula] and 2 planimetric methods [PM, Simpson's derived and length area methods]). This study showed that the GM measurement method tended to overestimate ESVI values. However, ESVI was closely related to the severity of heart failure in the MMVD group. Recent studies have shown that EDVI and ESVI measured by the GM method tended to be overestimated compared to actual values, and their diagnostic value in MMVD was significantly lower.

If left ventricular enlargement is determined solely by left ventricular diameter, many errors can occur. The **LVIDd/Ao ratio**, which divides the left ventricular end-diastolic diameter measured in 2D and M-mode by the aortic diameter, is a more useful indicator for assessing the degree of left ventricular enlargement. However, recent studies have shown that the LVIDd/Ao ratio is an echocardiographic parameter with high specificity but low sensitivity for differentiating asymptomatic patients or symptomatic from asymptomatic patients.

When the left ventricle enlarges, the **sphericity of the LV** increases (the left apex be-

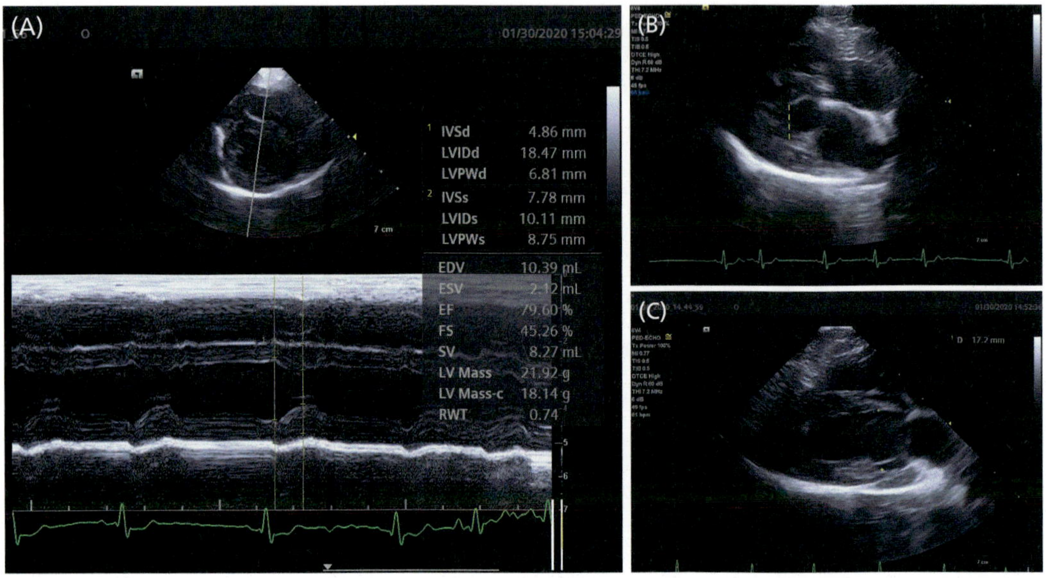

Figure 3-18 ▸ Left ventricular internal dimension measurement. (A) Measurement using M-mode echocardiography. (B and C) Measurement using 2D-echocardiography. (B) Systole, (C) Diastole.

comes more circular). One study confirmed a significant increase in LV sphericity in MMVD patients. The **LV sphericity index** can be calculated by dividing the **LV end-diastolic length (at left apical 2- or 4-chamber plane) on 2D (Figure 3-19A)** by the **M-mode LV end-diastolic diameter (Figure 3-19B)**.

Assessment of Myocardial Dysfunction

In some MMVD patients, left ventricular systolic function may be reduced. However, these left ventricular systolic function parameters vary greatly depending on the **LV loading condition**. Therefore, **fractional shortening (%FS)** and **LV ejection fractions (%LVEF)** are less accurate in assessing systolic function in MMVD patients. In some patient groups, they may even be increased due to compensatory responses, specifically due to dynamic left ventricular contraction resulting from increased preload and decreased afterload. Additionally, many Doppler parameters for evaluating LV function and filling pressure can be affected by age-related LV diastolic dysfunction, left atrial pressure overload due to MR, and changes in myocardial tissue velocities due to LV loading conditions. Therefore, echocardiographically measured left ventricular systolic and diastolic function can be misinterpreted.

In chronic MMVD patients, chronic volume overload causes morphological and functional changes in the left ventricle. The most commonly used left ventricular systolic function parameters are **%LVEF** and **%FS**:

- **LVEF = (end diastolic volume − end systolic volume)/end diastolic volume × 100**
- **FS% = (LVIDd − LVIDs)/LVIDd × 100**

%FS represents the percentage change in end-diastolic to end-systolic diameter. As

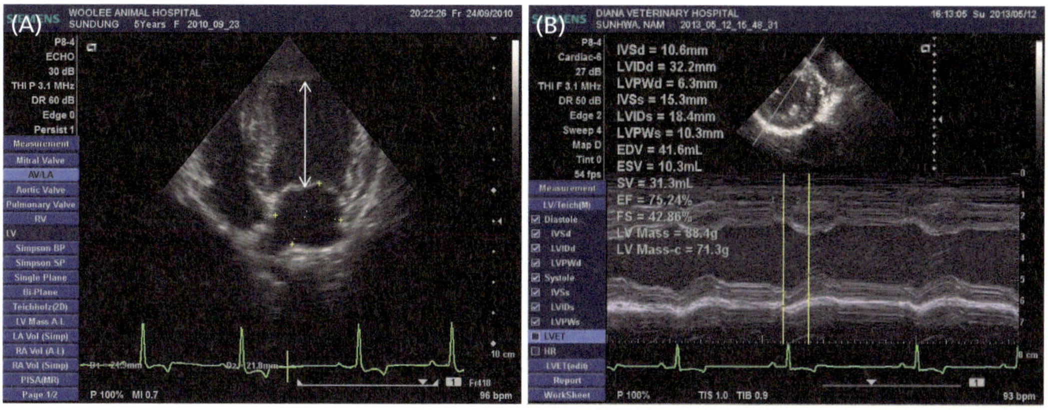

Figure 3-19 ▶ Left ventricular (LV) sphericity index is measured by dividing LV end-diastolic length (left apical 4-chamber view; A) by M-mode LV end-diastolic diameter (B).

mentioned previously, while %FS and %LVEF are reduced in humans with myocardial dysfunction, in MMVD patients, they are often increased due to dynamic left ventricular contraction resulting from increased preload and decreased afterload. Therefore, a normal or subnormal fractional shortening in the setting of severe MR and left ventricular dilation suggests systolic myocardial dysfunction, which is an indicator of disease progression.

Evaluation of Cardiac Function Using Spectral Doppler

To overcome the problems observed in simple left ventricular dysfunction measurements using 2D imaging, **Doppler and Tissue Doppler Imaging (TDI)** methods have been developed. When performing CW Doppler examination of mitral regurgitation (MR) flow, the MR flow velocity and flow profile are affected by left atrial pressure, left ventricular contractility, left heart system load, and systolic blood pressure (Figure 3-20A). Therefore, the peak velocity of MR is not well correlated with disease severity in the MMVD group. However,

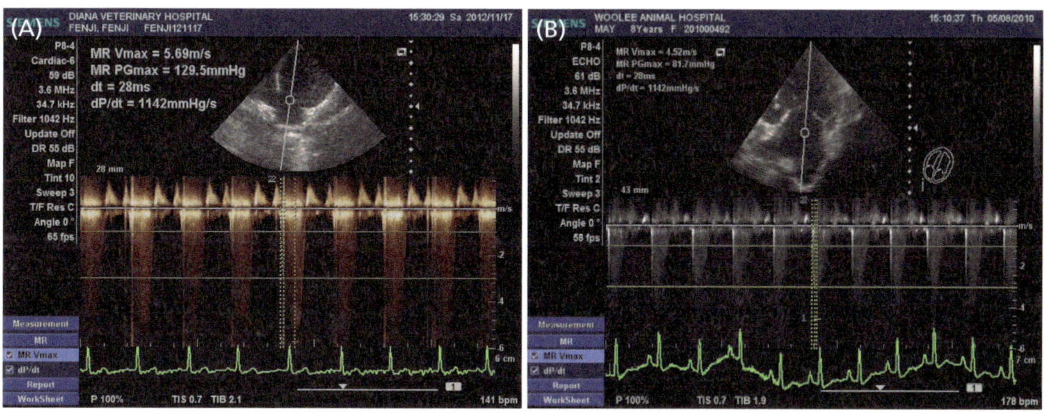

Figure 3-20 ▶ CW Doppler flow patterns in dogs with severe MR. (A) In this patient, MR peak velocity (pressure gradient) is 5.69 m/s (129.5 m/s). (B) Asymmetric MR flow profiles and cut-off signs (arrow) observed in patients with severe mitral insufficiency (see text).

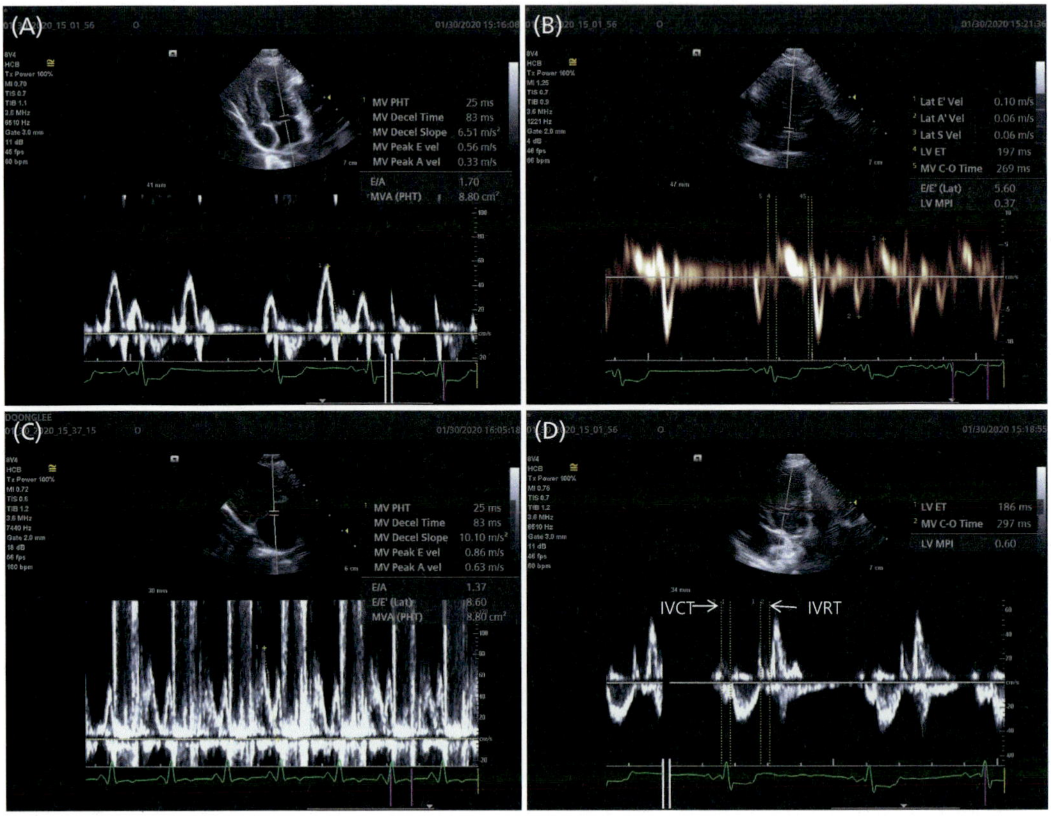

Figure 3-21 ▸ (A and B) Transmitral E/e ratio and E/IVRT (the isovolumic relaxation time), (B) tissue Doppler imaging of lateral e-peak (left apical 4-chamber plane), (C) E/E′ calculated based on Transmitral E-peak and TDI measured E′, (D) IVCT and IVRT can be measured by simultaneously sampling aortic flow and mitral flow.

asymmetric MR flow profiles and a **cut-off sign** observed in mid-to-late systole in patients with end-stage heart failure indicate advanced heart failure and are useful for differentiating ACVIM C and D groups (this phenomenon occurs due to reduced MR volume due to elevated left atrial pressure; Figure 3-20B). **dP/dt** represents the rate of pressure change during isovolumic contraction and is closely related to left ventricular systolic function in humans (Figure 3-20A). However, in most MMVD patients, canine dP/dt shows high peak velocity and asymmetric MR flow patterns, thus demonstrating limited diagnostic value.

PW Doppler findings for mitral inflow are closely related to disease severity in the MMVD group. Mitral inflow consists of an **E-peak** formed during rapid ventricular filling and an **A-peak** formed during atrial systole, and is closely related to left ventricular diastolic function (e.g., left ventricular relaxation ability, left ventricular volume overload/pressure overload, and recoil) and left ventricular filling pressure (Figure 3-21A). Although **E/A reversal (E/A <1)** is common in the elderly population, generally the E-peak is higher than the A-peak (i.e., E/A <1 is abnormal). In one study of 100 healthy dogs, the E-peak was 0.87±0.13 m/s and the A-peak was 0.61±0.12 m/s. High E-peak velocity (>1.5 m/s) indi-

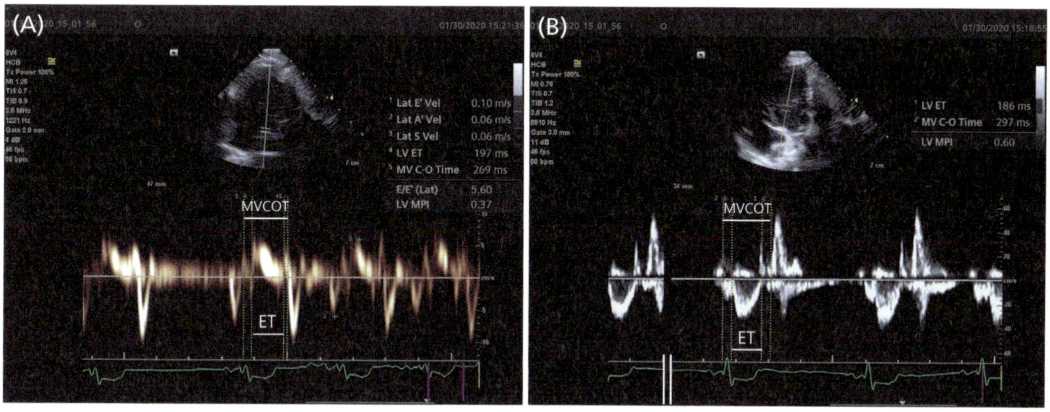

Figure 3-22 ▶ Left ventricular (LV) Tei index. (A) TDI measurement at the mitral annulus. (B) The Tei index can be calculated by simultaneously sampling aortic and mitral flow with pulse-wave Doppler (measurement method shown in Figure 3-21D). Detailed explanation in the text.

cates high left atrial pressure and was closely related to disease severity and survival rate in the MMVD group.

Generally, a high E-peak and short **E deceleration time** (<80 ms, in dogs over 10-12 years) indicate severe left atrial pressure overload and a left ventricle with reduced compliance. Conversely, an **E/A ratio <1 ± prolonged E deceleration time** indicates ventricular diastolic dysfunction. E-peak velocity increases with increasing left atrial pressure but decreases with left ventricular diastolic dysfunction. Therefore, mitral inflow velocity can sometimes lead to misinterpretation of left ventricular dysfunction.

However, since myocardial and mitral annular velocities are less affected by intracardiac loading conditions, **E/e'** (mitral annular velocity measured by TDI) better reflects left ventricular filling pressure (left atrial pressure) in MMVD patients. A decrease in **e'-peak** at the mitral annulus indicates ventricular diastolic dysfunction. One study showed that an **E/e' ratio >9** indicated a left atrial pressure >20 mmHg, and **E/e' >12** (with high E-peak velocity) indicated a state of congestive heart failure. Another study showed that **E/e' >13** had 80% sensitivity and 83% specificity for diagnosing congestive heart failure. Although TDI e'-peak is less affected by left ventricular loading conditions, the E/e' ratio has limited diagnostic value in patients with these conditions because it can be affected by age and severe left ventricular volume overload. Several studies have shown that **E-peak/the isovolumic relaxation time (E/IVRT)** was closely related to left atrial pressure and left ventricular end-diastolic pressure (Figure 3-21C and D). IVRT can decrease as left ventricular filling pressure increases. One study showed that an **E/IVRT ratio >2.5** and **IVRT <45 ms** indicated a state of congestive heart failure. Evaluation of pulmonary venous flow with PW Doppler could estimate the degree of left atrial volume overload and pressure overload in humans. However, pulmonary veins are located in the far field on ultrasound, which makes accurate evaluation difficult. **LV Tei index (LV Tei = [IVRT+IVCT+LVET]/LVET, IVCT for the isovolumic contraction time)** is a useful indicator in humans for simultaneously assessing left ventricular systolic and diastolic dysfunction (Figure 3-22). Human studies showed that a shortened LVET and increased LV Tei index indicated systolic dysfunction, and an increased mitral valve closure time and Tei index indicated diastolic dysfunction. In studies on canine MMVD patients, the LV Tei index was closely related to LV dP/dt and disease severity. However, in moderate-to-severe MR patients, IVRT and IVCT are often too short or undetectable. Therefore, the Tei index is rarely used in veterinary medicine.

Evaluation of Pulmonary Hypertension

Pulmonary venous hypertension (PVH) is one of the common causes of pulmonary hypertension in dogs and is prevalent in MMVD patients (especially in advanced heart failure). In dogs, standard methods for detecting this pulmonary hypertension include measuring **tricuspid regurgitation (TR) velocity** and/or **pulmonic regurgitation (PR) velocity**. One study demonstrated that pulmonary hypertension was more prevalent in more advanced MMVD patients through Doppler-based evaluation. Another study showed that MMVD patients with moderate to severe pulmonary hypertension had a poorer prognosis.

TR and PR Jet Velocity Measurement

The **TR jet velocity** measured from the right parasternal short-axis view or left apical 4-chamber long-axis view indicates the **systolic pulmonary artery pressure (sPAP)** in the absence of right ventricular outflow tract obstruction. sPAP can be calculated by measuring the peak TR jet velocity using the **modified Bernoulli equation ($\Delta P = 4 \times$ velocity2)** (Figure 3-23A), and **mean pulmonary artery pressure (mPAP)** can be calculated by measuring the peak PR jet velocity (Figure 3-23B). The calculated pressure gradient must then be added to the **right atrial pressure (RAP)** ($\Delta P + \pi$ right atrium). RAP is estimated as follows: 5 mmHg if there is no right atrial enlargement, 10 mmHg if there are signs of right atrial enlargement, and 15 mmHg if there are symptoms of right heart failure (e.g., ascites). mPAP can also be calculated by the following formula: **mPAP = 0.61 × sPAP + 2**. In patients with

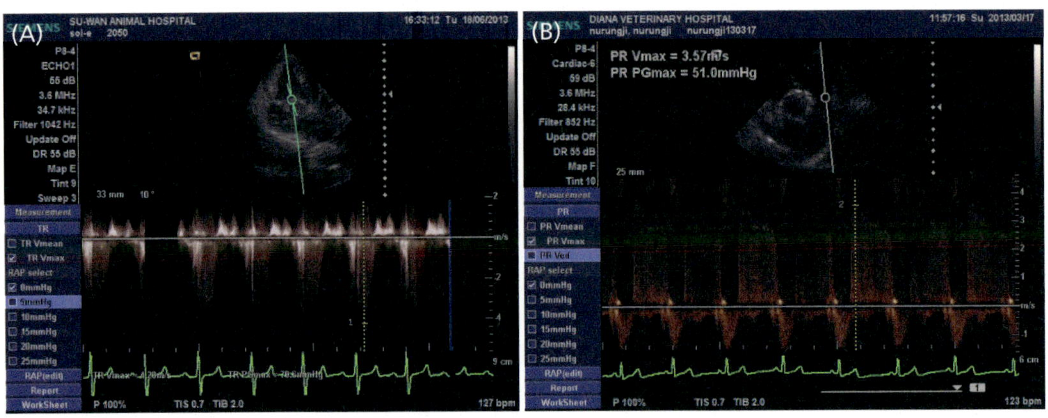

Figure 3-23 ▸ Echocardiographic findings of pulmonary hypertension. (A) Severe tricuspid regurgitation measured by CW Doppler. (A) Severe pulmonic regurgitation measured by CW Doppler. More detailed explanation in the text.

only PR, mPAP can be calculated by the following formula: **PR peak pressure gradient + RAP**. mPAP <25 mmHg is considered normal, mPAP 25-40 mmHg (equivalent to 2.8-3.4 m/s of peak TR velocity) is mild, mPAP 41-55 mmHg (3.4-4.3 m/s) is moderate, and >55 mmHg (>4.3 m/s) is severe. However, if the ultrasound beam is not parallel to the direction of blood flow, the actual severity of pulmonary hypertension may be underestimated. Another study showed that SAP calculated by the Doppler method tended to overestimate the actual severity of pulmonary hypertension in MMVD patients. In some patients with pulmonary hypertension, TR or PR may not be observed at all on Doppler examination. Therefore, in such patients (if they have symptoms of right heart failure or persistent coughing), other echocardiographic parameters for pulmonary hypertension should be examined. Despite these limitations, SAP measured in MMVD patients was an excellent prognostic factor for predicting the progression of congestive heart failure and survival time.

2D-Echocardiographic Findings Indicating Pulmonary Artery/Right Ventricular Outflow Tract Dilation

In MMVD patients, 2D-Echocardiographic findings indicating dilation of the pulmonary artery or right ventricular outflow tract provide supportive evidence of pulmonary hypertension, although pulmonic stenosis without pulmonary hypertension may also cause post-stenotic pulmonary artery dilation. Chronic volume overload and/or pressure overload in the right ventricle can cause concentric and/or eccentric hypertrophy, so measuring the right ventricular wall thickness or internal dimension using 2D echocardiography can help confirm pulmonary hypertension. Furthermore, **left ventricular eccentricity** (expressed as the ratio of the width to the height of the lumen in the left ventricular short-axis view; Figure 3-24C) can easily identify the presence of pulmonary hypertension. Two studies reported that when right ventricular pressure exceeded left ventricular pressure, LV eccentricity became less than 1. This finding is commonly observed in patients with pulmonary hypertension.

In normal dogs, the diameter of the **main pulmonary artery (MPA)** is almost equal to the diameter of the aorta. Therefore, ideally, the **MPA/Ao ratio** should be close to 1 (Figure 3-24D). Consequently, an **MPA/Ao ratio >1** indicates pulmonary artery dilation and the presence of pulmonary hypertension (assuming no right ventricular outflow tract obstruction). Many studies have confirmed MPA dilation in dogs with pulmonary hypertension.

The **right pulmonary artery distensibility index (RPADi)** is a method of measuring the diameter of the pulmonary artery cross-section during systole and diastole from M-mode pulmonary artery short-axis view (Figure 3-24A and B) or 2D left apical long-axis view (Figure 3-25A). One study showed that **RPADi <35** suggested pulmonary hypertension, with **RPADi 28-35** indicating mild, **23-27** moderate, and **<23** severe pulmonary hypertension. Additional studies showed that RPADi was closely correlated with systolic pulmonary pressure. However, the reference range for RPADi varied greatly depending on the ultrasound plane.

RVOT-%FS is an index of right ventricular systolic function that can be calculated from M-mode measurements in the aortic short-axis view using the following formula: **([RVOT**

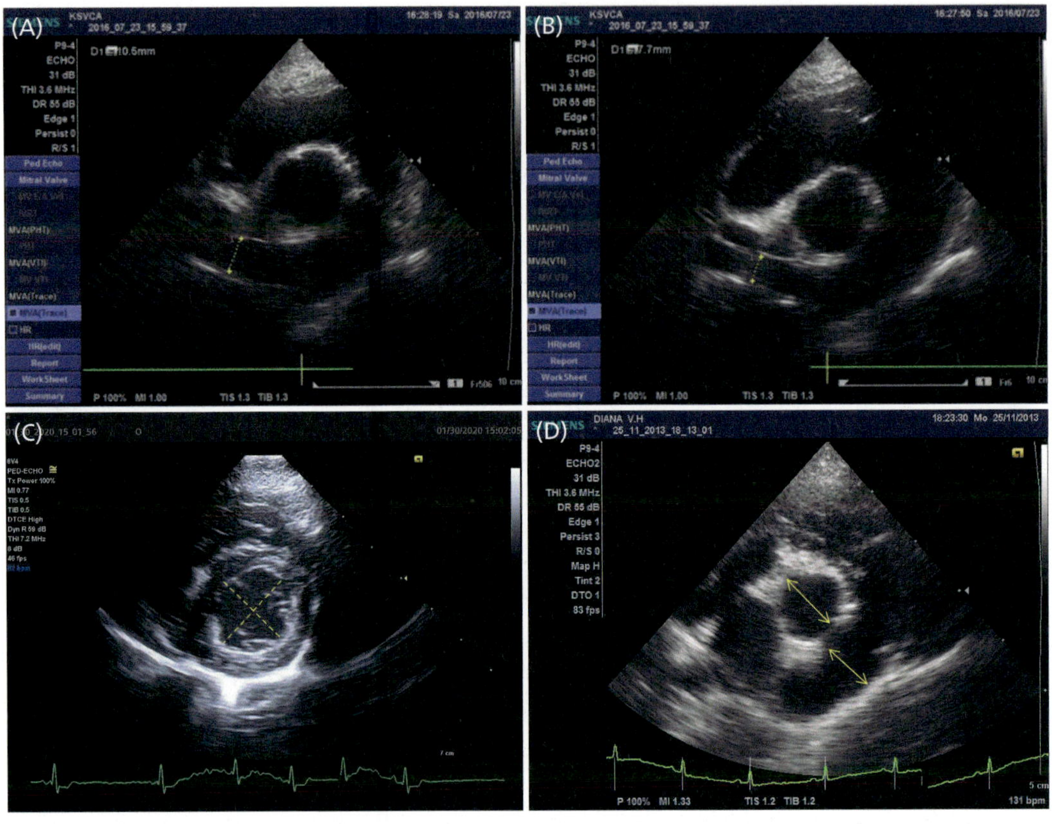

Figure 3-24 ▶ 2D-echocardiography for pulmonary hypertension. (A and B) Right pulmonary artery distensibility index (RPADi). (C) Left ventricular (LV) eccentricity. (D) Ratio of main pulmonary artery diameter to aortic root diameter. More detailed explanation in the text.

dimensions at end diastole - RVOT dimensions at end systole]/RVOT) × 100 (Figure 3-25B). One study showed that RVOT-%FS was significantly reduced (<45%) in the presence of pulmonary hypertension. However, RVOT-%FS can be affected by other diseases that can cause right ventricular systolic dysfunction.

Tricuspid annular plane systolic excursion (TAPSE) also indicates right ventricular systolic function and is highly influenced by the right ventricular pressure load. Recent canine studies have confirmed low TAPSE values in dogs with pulmonary hypertension. TAPSE can be calculated by horizontally placing the cursor at the lateral tricuspid annulus in the left apical 4-chamber long-axis view on M-mode and measuring the maximal and minimal heights (Figure 3-25C). However, TAPSE varies greatly with patient size, so recent studies have used **normalized TAPSE (nTAPSE = TAPSE/[BW(kg)]1/3)** or **TAPSE/Ao ratio** to minimize this effect. The median (range) of nTAPSE was 0.60 (0.53-0.66) for ACVIM B1, 0.71 (0.64-0.84) for B2 group, and 0.73 (0.58-0.80) for C and D groups. Another study showed that the TAPSE/Ao ratio was closely related to left atrial size and the severity of pulmonary hypertension in the MMVD group. **TAPSE/Ao ratio <0.65** suggested the presence of pulmonary hypertension in the MMVD group. However, the diagnostic sensitivity was

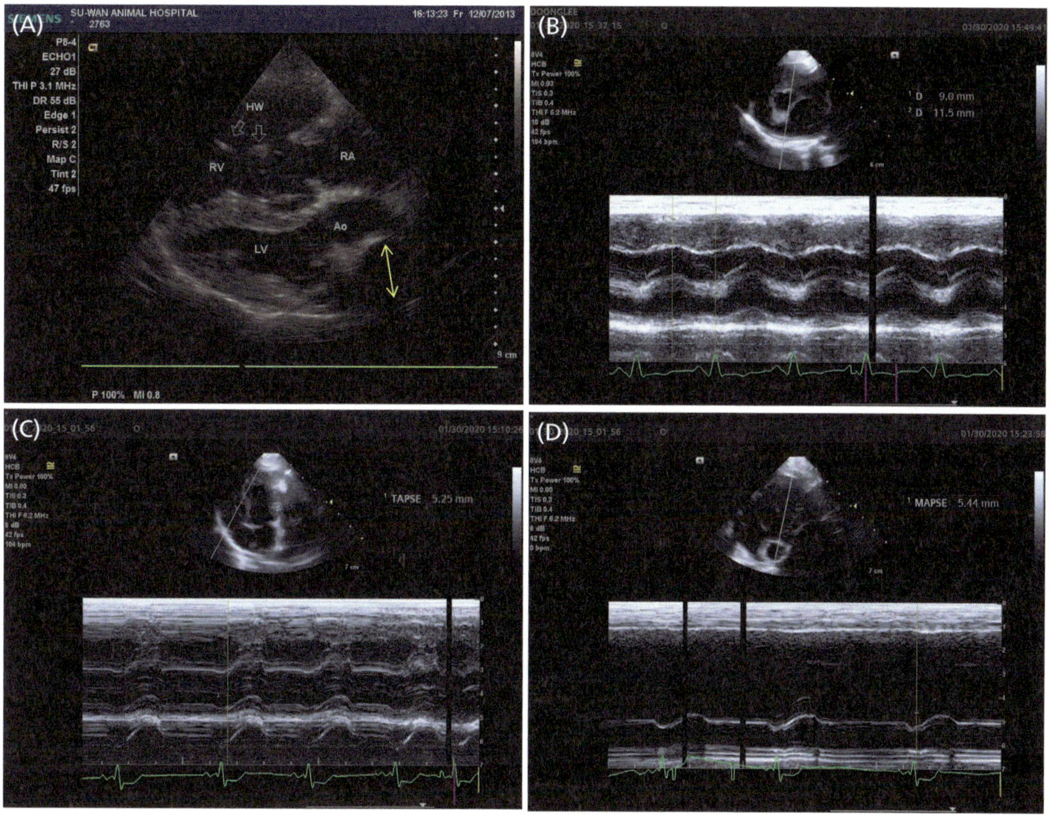

Figure 3-25 ▸ 2D and M-mode echocardiographic examination for pulmonary hypertension. (A) Right pulmonary artery distensibility index (RPADi) measured from the right parasternal long-axis view (arrow). (B) Right ventricular outflow tract fractional shortening (RVOT-%FS). (C) Tricuspid annular plane systolic excursion (TAPSE). (D) Mitral annular plane systolic excursion (MAPSE). More detailed explanation in the text.

not high. However, the interpretation of TAPSE requires contextualization. In patients with certain congenital diseases (such as severe pulmonic stenosis) characterized by marked concentric hypertrophy, TAPSE may be low, thereby underestimating the true right ventricular systolic function. Conversely, in patients with Myxomatous mitral valve disease (MMVD), the volume overload and hyperdynamic state of the left ventricle can lead to an increase in TAPSE. The TAPSE/pulmonary artery systolic pressure (PASP) ratio has been proposed as a noninvasive index of right ventriculo-arterial coupling. Measures of right ventricular systolic function (like TAPSE) are known to be affected by other diseases that cause systolic dysfunction.

Doppler Flow Patterns and Related Indicators of Right Ventricular Outflow Tract (Pulmonary Artery)

Various quantitative echocardiographic parameters for pulmonary hypertension have been evaluated in humans and dogs. These parameters include **PW Doppler pulmonary artery flow acceleration time (AT)**, **AT to the ejection time of PA flow ratio (AT/ET)** (Figure

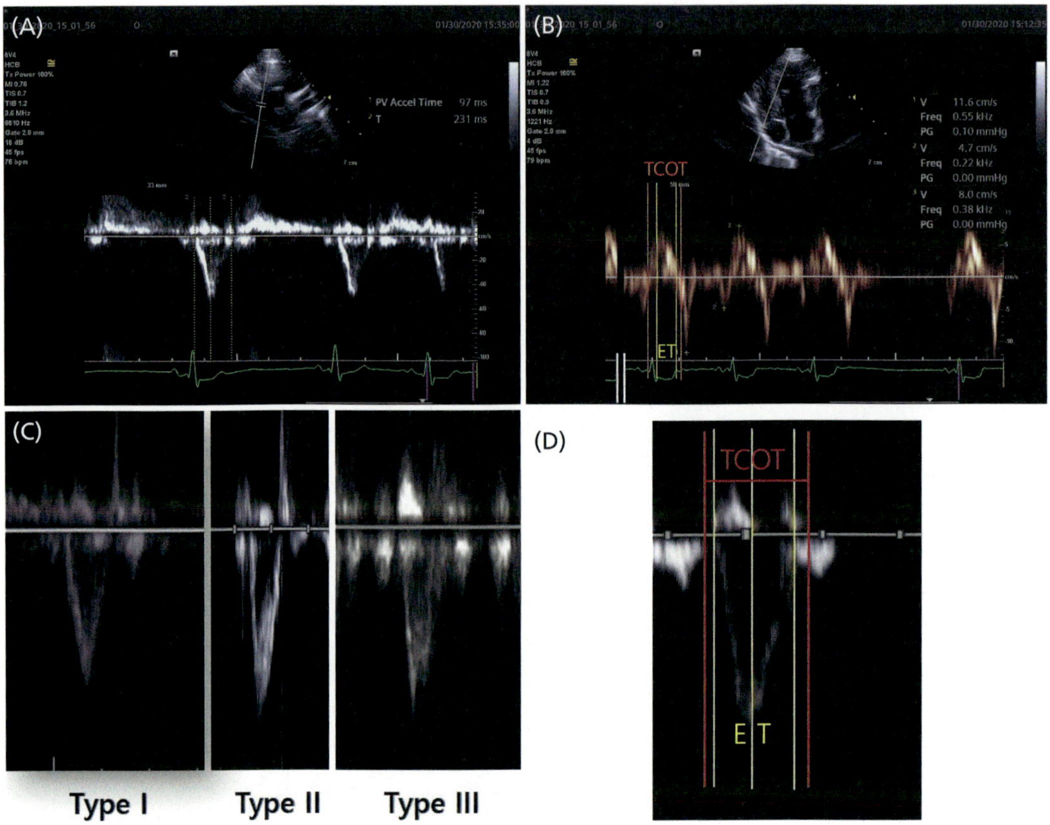

Figure 3-26 ▸ PW Doppler and TDI findings for pulmonary hypertension. (A) PW Doppler-derived acceleration time (AT) to peak pulmonary artery (PA) flow velocity (AT), AT to the ejection time (ET) of PA flow ratio (AT/ET). (B) Tissue Doppler-derived right ventricular Tei index (RV Tei). (C) PW Doppler-derived pulmonic outflow profiles. (D) PW Doppler–derived RV Tei index. More detailed explanation in the text.

3-26A), **right ventricular Tei index (RV Tei)** (Figure 3-26B and D), and **Doppler pulmonary artery flow pattern** (Figure 3-26C). One study confirmed that AT, AT/ET ratio, and RV Tei index were closely related to systolic pulmonary artery pressure in dogs. In particular, an **AT/ET ratio ≤ 0.25** strongly suggests pulmonary hypertension, while an **AT/ET ratio ≥ 0.42** can rule out pulmonary hypertension. However, other studies found a strong correlation between Doppler-calculated mean pulmonary artery pressure and AT and AT/ET, but a weak correlation with RV Tei index. In this study, over 50% of ISACHC II and III MMVD patients showed ambiguous AT/ET values between 0.25-0.42. Therefore, this indicator had low diagnostic sensitivity. However, most individuals with TR or PR showed an AT/ET ≤ 0.25, indicating high specificity for diagnosing clinically significant pulmonary hypertension. Furthermore, in this study, the Doppler pulmonary artery flow pattern was closely related to the severity of pulmonary hypertension. However, these flow patterns can vary depending on the degree of right ventricular load and systolic function, which may reduce their diagnostic value in individuals with dynamic right ventricular contraction. Although right ventricular TDI examination cannot directly diagnose pulmonary hypertension, it can be useful in evaluating right ventricular systolic and diastolic dysfunction in these patients.

One study measured myocardial velocities of the right ventricular free wall during systole (Stdi), early diastole (Etdi), and late diastole (Atdi) to calculate a **global TDI index**: **global TDI index = Stdi × Etdi/Atdi**. The results showed that the cut-off values for diagnosing pulmonary hypertension were **global TDI <11.8 (sensitivity 89%, specificity 90%)** and **Etdi/Atdi <1.12 (sensitivity 89%, specificity 93%)**.

Measurement of Cardiac Chamber Deformation and Systolic Dyssynchrony

Strain and strain rate imaging are methods that assess the degree of myocardial deformation by dividing the myocardium into segments. **Myocardial strain** (i.e., the percentage of myocardial segment deformation from its original shape) and **strain rate** (myocardial deformation velocity; S-1) can be used to evaluate intrinsic myocardial function and are less angle-dependent than TDI. Strain and strain rate can be measured in the left ventricle, left atrium, and right ventricle using **speckle tracking echocardiography (STE)**, and can be measured by dividing the cardiac chambers into longitudinal, radial, and circumferential planes (Figure 3-27). STE assesses the direction, velocity, and extent of myocardial deformation frame by frame continuously throughout the cardiac cycle by assigning acoustic markers (speckles) to myocardial segments.

Although evaluating systolic function is important in MMVD patients for drug treatment and progression assessment, this assessment in conventional echocardiography can vary greatly depending on the **cardiac loading conditions**. Several studies have shown that evaluating strain and strain rate using **2D-STE** can be useful in predicting disease progression in MMVD patients. However, one study using GE analysis software (EchoPAC) to evaluate **longitudinal strain** showed poor reproducibility and consistency of results, while **radial strain** measured in the short-axis view was more consistent and reproducible. This strain analysis uses automated software programs, so special attention is needed for the following points before examination: (1) accurately selecting the start and end points of the cardiac cycle while observing the ECG, (2) precisely marking the endocardial region for automated analysis, and (3) ensuring that the region of interest is well included. For detailed examination methods, refer to Smith et al. **Velocity vector imaging (VVI)** is a strain analysis program used in Siemens ultrasound equipment that can evaluate the direction and intensity of myocardial deformation using vectors (Figure 3-27 C and D).

Atrial myocardial deformation can also be measured using TDI and 2D-STE and has recently been used in human and canine studies to measure the amount of load on the left atrium. Although these methods showed several problems (e.g., suboptimal reproducibility, angle dependence, and confounding effect of noise artifacts), they were more useful indicators than evaluating left atrial function using conventional 2D imaging. Specific STE parameters related to left atrial function include **peak atrial longitudinal average strain (PALS), peak atrial contraction average strain, and contraction strain index (CSI)**. These indicators reflect the left atrial reservoir function during left ventricular filling (PALS), the left atrial pump function (PACS), and the left atrial active contraction function (CSI) (Figure

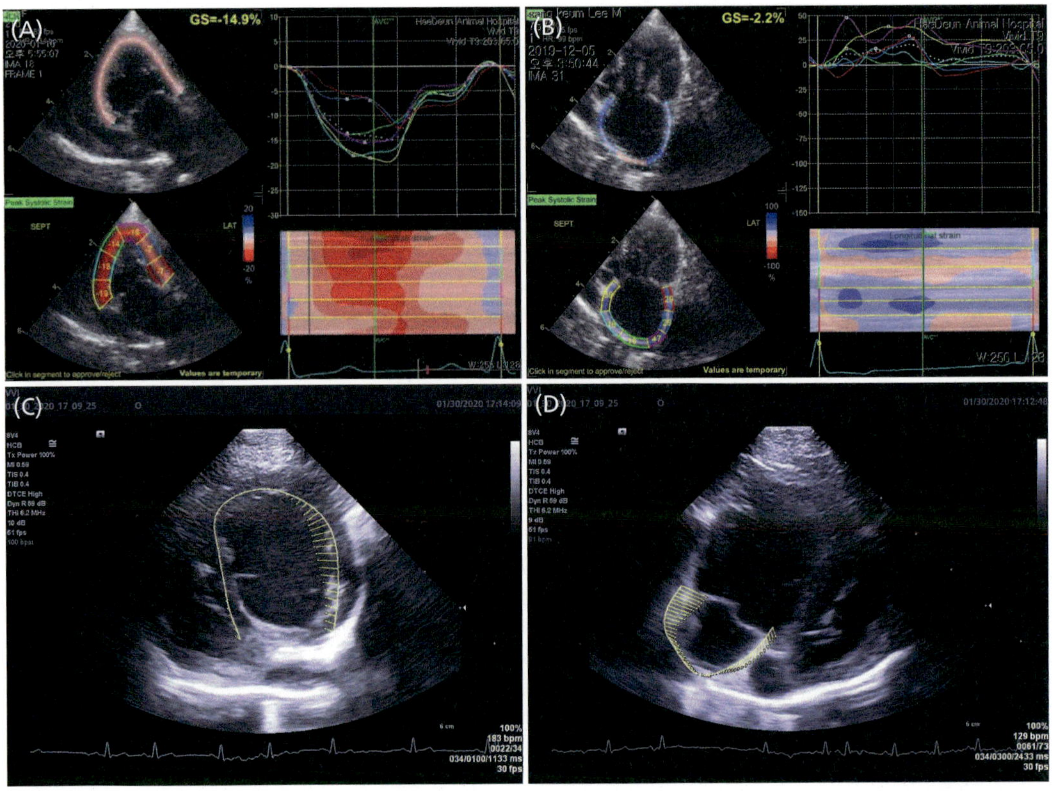

Figure 3-27 ▸ Strain and strain rate of the left atrium and left ventricle measured in MMVD dogs. (A and B) STE of the left ventricle (A) and left atrium (B): Measured with GE analysis software algorithm (EchoPAC). (C and D) STE of the left ventricle (C) and left atrium (D): Measured with Siemens vector velocity imaging (VVI). More detailed explanation in the text.

3-27B). In humans, left atrial strain analysis was useful in assessing disease severity in patients with valvular disease, atrial fibrillation, and acute coronary artery disease. However, a recent study on left atrial STE parameters in canine MMVD patients found that they were not useful for differentiating ACVIM B1 and B2 groups but were only useful for differentiating ACVIM B2 and C groups. **PALS <27.9%** showed 100% diagnostic sensitivity and specificity for differentiating ACVIM B2 and C groups. Another study also showed that STE parameters such as PALS, PACS, and CSI were useful in evaluating MMVD progression. Furthermore, **PALS <30% and CSI per 1% increase** were useful for predicting cardiac-related mortality.

The right ventricle is crescent-shaped and surrounded by the left ventricle, making it often difficult to accurately assess right ventricular function using echocardiography. Human studies have shown that TDI and STE methods can overcome these limitations. Recent canine studies on healthy dogs have also shown that these examination methods are applicable and reproducible. Furthermore, other studies have shown that the MMVD group had significantly different **RV longitudinal strain** and **dyssynchrony index** compared to the normal control group. In this study, **RV longitudinal strain** was significantly lower and **sys-**

tolic shortening time strain was slower in the group with type 2 pulmonary hypertension (e.g., MMVD) compared to the control group.

3.9 Diagnostic Guideline for Determining the Severity of Mitral Valve Disease

A. The Mitral INsufficiency Echocardiographic (MINE) score: A severity classification of myxomatous mitral valve disease in dogs (JVIM 35, 1238–1244)

Background

- Myxomatous mitral valve disease (MMVD) is the most common acquired heart disease in dogs (~75% of cases).
- Clinical classification typically follows ACVIM guidelines, but **no simple, universally accepted echocardiographic severity score** exists.
- Echocardiography is essential to assess cardiac remodeling, predict CHF, decide treatment timing, and evaluate prognosis.
- The study proposed the **MINE score**, a simple echocardiographic system based on routine variables.

Methods

- **Design:** Retrospective, multicenter, observational study.
- **Population:** 560 dogs with MMVD (2011–2019, Italy).
- **Exclusions:** <3 years, >20 kg, FS <25%, atrial fibrillation, pregnancy, or other systemic/cardiac diseases.
- **Echocardiographic variables in the MINE score:**

  1. **LA/Ao ratio** (left atrium-to-aorta ratio)
  2. **LVIDDn** (LV end-diastolic diameter normalized for BW)
  3. **Fractional shortening (FS%)**
  4. **E-wave transmitral peak velocity (E-vel)**

- **Scoring:**

  - Mild = 4–5 points
  - Moderate = 6–7 points
  - Severe = 8–12 points
  - Late stage = 13–14 points

Results

- Distribution: 43% mild, 20% moderate, 33% severe, 4% late stage.
- **Median survival times (cardiac death):**
  - Mild: 2344 days (~6.5 years)
  - Moderate: 1882 days (~5 years)
  - Severe: 623 days (~1.7 years)
  - Late stage: 157 days (~5 months)
- **Prognostic power:**
  - A score >8 strongly predicted cardiac death (AUC 0.85, sensitivity 87%, specificity 73%).
  - Each variable (LA/Ao, LVIDDn, FS, E-vel) was independently predictive of outcome.

Box 1. The Mitral INsufficiency Echocardiographic (MINE) score used four echocardiographic variables, all routinely obtainable in clinical practice:

a) Left atrium-to-aorta ratio (LA/Ao)

- Method:
  - Measured in right parasternal short-axis view at the level of the aortic valve.
  - Ratio of the left atrial diameter to the aortic root diameter.
- Rationale: Reflects left atrial enlargement, a hallmark of chronic volume overload from MR.
- Scoring cutoffs:
  - <1.70 → score 1
  - 1.70–1.90 → score 2
  - 1.91–2.50 → score 3
  - 2.50 → score 4

b) Left ventricular end-diastolic diameter normalized to body weight (LVIDDn)

- Method:
  - Measured in M-mode from right parasternal short–axis view at papillary muscle level.
  - Normalized to body weight (using allometric scaling).
- Rationale: Quantifies left ventricular dilation due to chronic MR.
- Scoring cutoffs:
  - <1.70 → score 1
  - 1.70–2.00 → score 2
  - 2.10–2.30 → score 3
  - 2.30 → score 4

c) Fractional Shortening (FS%)

- Method:
  - Derived from M-mode (right parasternal short-axis view).
  - Formula: FS% = [(LVIDd – LVIDs) / LVIDd] × 100
- Rationale: Represents systolic function; values often supranormal in MR due to volume overload.
- Scoring cutoffs:
  - <45% → score 1
  - 45–50% → score 2
  - 50% → score 3

d) E-wave transmitral peak velocity (E-vel)

- Method:
  - Obtained via pulsed–wave Doppler at the tips of mitral valve leaflets from left apical 4–chamber view.
  - Measures peak velocity of early diastolic transmitral inflow (E–wave).
- Rationale: Reflects LV filling pressure and severity of MR.
- Scoring cutoffs:
  - <1.20 m/s → score 1
  - 1.20–1.50 m/s → score 2
  - 1.50 m/s → score 3

- MINE score refined prognosis even within the same ACVIM stage (e.g., distinguishing faster vs. slower progressors).

Clinical Significance

- The **MINE score is easy-to-use, reproducible, and clinically meaningful.**
- It offers prognostic information beyond ACVIM staging and can guide:
  - Frequency of follow-up
  - Treatment decisions
  - Identifying high-risk asymptomatic dogs
  - Potential candidates for mitral valve surgery
- Useful for both **clinical management** and **research standardization.**

Reference:

Vezzosi T, Grosso G, Tognetti R, Meucci V, Patata V, Marchesotti F, Domenech O.

The Mitral INsufficiency Echocardiographic score: A severity classification of myxomatous mitral valve disease in dogs. *J Vet Intern Med*. 2021;35(3):1238-1244.

B. Efficacy of a mitral regurgitation severity index (MRSI) to predict long-term outcome in dogs with MMVD (JVIM 38, 51–60)

Background

- MMVD progression varies widely, even after cardiomegaly develops (ACVIM Stage B2).
- Previous prognostic indices (e.g., MINE score) are helpful but rely on advanced echocardiography.
- The **Mitral Regurgitation Severity Index (MRSI)** was developed as a **simpler formula:**

$$MRSI = (HR/120) \times (LA/Ao) \times (Age/10) \times 100$$

- This study tested its predictive power across large, diverse datasets.

Methods

- **Design:** Retrospective, pooled analysis of 4 prior study samples.
- **Population:** 869 dogs with MMVD (Stage B2 or C).
- **Data collected:** HR, LA/Ao, age, radiographic measures (VHS, VLAS, MVLAS), survival outcomes.
- **Alternative formulas:** replaced LA/Ao with radiographic indices (MRSI-VHS, MRSI-VLAS, MRSI-MVLAS).

Box 2. The Mitral Regurgitation Severity Index (MRSI) used a simple mathematical formula combining age, heart rate, and left atrial size:

a) Heart Rate (HR)

- Method:
  - Measured at rest, preferably during physical exam.
  - If not available, values taken from ECG or echocardiogram.
- Normalization: Divided by 120 bpm (approximate median in MMVD dogs).
- Rationale: Tachycardia is linked with sympathetic activation and worse outcomes.

b) Left Atrium-to-Aorta Ratio (LA/Ao)

- Method:
  Same as in MINE score (2D or M-mode echo, right parasternal short-axis).
- Rationale: Key indicator of left atrial enlargement due to MR.

c) Age

- Method:
  Dog's age in years, divided by 10 for scaling.
- Rationale: Older age is associated with poorer prognosis in MMVD.

d) MRSI Formula

- MRSI=(HR120)×(LA:Ao)×(Age10)×100\text{MRSI} = \left(\frac{HR}{120}\right) \times (LA:Ao) \times \left(\frac{Age}{10}\right) \times 100MRSI=(120HR)×(LA:Ao)×(10Age)×100
- Interpretation cutoffs:
  - 156 → shorter time to CHF onset
  - 173 → shorter survival

e) Alternative Radiographic Substitutes for LA/Ao

Since not all practices can perform echocardiography, the study also validated MRSI using thoracic radiographs:

1. Vertebral Heart Size (VHS)

- Long and short axes of the heart measured on lateral thoracic radiograph.
- Summed and compared to thoracic vertebral body lengths (starting at T4).

2. Vertebral Left Atrial Size (VLAS)

- A line from the dorsal junction of the caudal vena cava and left atrium to the most caudal border of the LA.
- Transposed onto vertebral column for measurement.

3. Modified VLAS (MVLAS)

- Similar to VLAS, but a second line drawn perpendicular to improve reproducibility.

- Formulas with substitution:
  - MRSI-VHS = (HR/120) × (VHS/10.5) × (Age/10) × 100
  - MRSI-VLAS = (HR/120) × (VLAS/2.5) × (Age/10) × 100
  - MRSI-MVLAS = (HR/120) × (MVLAS/3.4) × (Age/10) × 100

- Outcomes: **time to CHF onset/recurrence** and **all-cause death.**

Results

- **Stage B2 dogs (n=237):**
  - MRSI >156 predicted earlier CHF (median 407 vs 1404 days).
  - MRSI >173 predicted shorter survival (868 vs 1843 days).
  - Hazard ratio for CHF onset = 3.02; for death = 4.26.
  - MRSI had **better predictive ability than individual variables (HR, age, LA/Ao).**
  - Comparable prognostic accuracy to the MINE score.
  - Radiographic versions (MRSI-VLAS, MRSI-MVLAS) performed similarly, making the tool accessible to general practice.

- **Stage C dogs (n=670):**
  - MRSI was **not consistently predictive.**
  - Removing age from the formula improved prognostic performance in some subsets.

- **Treatment effects:** Pimobendan, ACE inhibitors, or spironolactone use did **not significantly alter MRSI's predictive value.**

Clinical Significance

- **MRSI is a practical prognostic tool for Stage B2 MMVD dogs.**
- Advantages:
  - Simple calculation
  - Requires only HR, age, and LA/Ao (or radiographic substitutes)
  - Can be used in primary practice without advanced Doppler echo

- Not as reliable for Stage C, but still informative in earlier disease.
- Helps guide **owner counseling, recheck scheduling, and risk stratification.**

Note:

- **MINE score**
  - Echocardiography-based, detailed, stronger for research and referral settings.
  - Purely echocardiographic, uses systolic and diastolic function + chamber size.

- **MRSI**
  - Simpler, can substitute radiographs for echo, practical for general practice.
  - Simple index using HR, age, and LA size (echo or X-ray), easier to apply in general practice.

- Both scores improve prognostication beyond ACVIM staging, especially in Stage B2 (preclinical enlargement).

Reference:

Vereb M, Atkins CE, Adin D, Blondel T, Coffman M, Lee S, Guillot E, Ward JL. Efficacy of a mitral regurgitation severity index to predict long-term outcome in dogs with myxomatous mitral valve disease. *J Vet Intern Med.* 2024;38(1):51-60.

CANINE MYXOMATOUS MITRAL VALVE DISEASE

CHAPTER 04

Case Studies

CASE 1

Patient Information

A 9-year-old female Maltese (body weight 3.2 kg).

History and Physical Examination

The patient presented to a local hospital with dyspnea, syncope, tachypnea, cyanosis, and coughing. She had been prescribed diltiazem (1 mg/kg tid), furosemide (1 mg/kg bid), pimobendan (0.25 mg/kg bid), and ramipril (0.125 mg/kg sid). The owner reported that the patient had been making a goose-like sound since spring and her symptoms improved somewhat after taking a bronchodilator. She frequently coughed (especially at night and in the morning) and had even fainted due to severe coughing. On physical examination, a holosystolic murmur (grade V/VI) was auscultated at the left apical region of the heart (Figure 4-1). Her systolic blood pressure was 150 mmHg, heart rate 153 bpm, and respiratory rate 35 breaths/minute.

Diagnostic Tests

Laboratory Evaluation: No specific findings were observed on complete blood count. Mild hyperglycemia and elevated BUN levels were noted, but no other significant abnormalities were present (Tables 4-1, 4-2).

ECG Examination: Heart rate was 180-190 bpm, and the rhythm itself was sinus. The

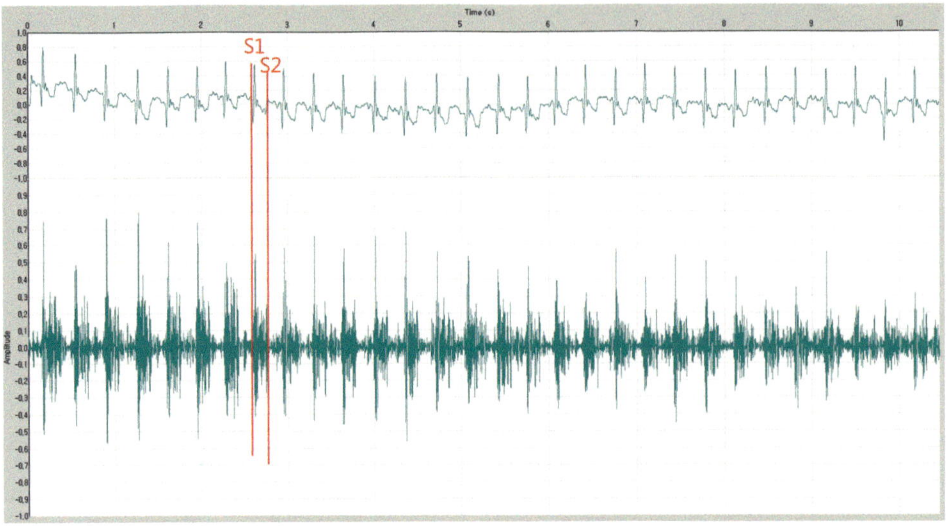

Figure 4-1 ▶ Phonocardiogram findings recorded from the patient's left apical region. A murmur is observed only during systole, and its shape is a holosystolic plateau-shaped regurgitant murmur.

QRS amplitude was ~2.0 mV, and QRS duration was 0.05-0.06 sec, slightly suggesting left ventricular enlargement. P-mitrale indicated left atrial enlargement (Figure 4-2). The heart rhythm was mostly regularly regular, and no specific arrhythmias were observed.

Radiographic Examination: Ventrodorsal thoracic radiographs showed left atrial enlargement (arrow) and caudal pulmonary vein distention, but no pulmonary infiltrates were observed (Figure 4-3A). Lateral thoracic radiographs revealed tracheal elevation, severe left atrial enlargement (arrow), main bronchial compression, cardiomegaly (VHS = 11V, L = 6.5V, S = 4.5V), and apical rounding (Figure 4-3B). These findings are typical for left cardiomegaly. The patient also had severe tracheal collapse.

Echocardiographic Examination: In the B-mode, left apical two-chamber long-axis view, distinctly thickened and club-shaped mitral valve leaflets were identified (Figure 4-4A pm Video Material 4-1). Mitral valve prolapse with coaptation protruding into the

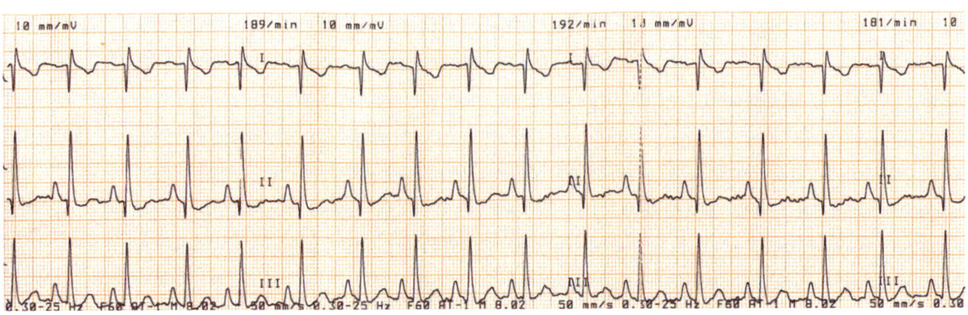

Figure 4-2 ▶ Patient's electrocardiogram findings.

Table 4-1 ▸ Patient's complete blood count findings

Test Item	Reference Range	On Admission	One Month Later	Two Months Later	Three Months Later
WBC	6–17 K/uL	14.88	18.60	15.07	13.20
RBC	5.5–8.5 M/uL	8.28	7.22	6.99	6.85
HEMOGLOBIN	12–18 g/dL	18.9	17.0	15.3	15.9
MCV	60–77 fL	64	65	65	59.8
MCH	19.5–24.5 PG	22.8	23.5	21.9	23.2
MCHC	31–34 g/dL	35.8	36	33.7	38.8
PLATLETS	200–500 k/uL	469	618	501	356
LYMPHS(ABSOLUTE)	1–4.8 k/uL	1.10	0.51	0.45	0.67
MONOS(ABSOLUTE)	0.2–1.5 k/uL	0.88	1.01	1.12	0.49
GRA(ABSOLUTE)	3–12	12.90	17.07	13.50	
RDW	12–18%	16.2	15.5	17.1	13.5
LYMPHO	12–30%	7.4	2.7	3.0	0.67
MONOCYTE	2–4%	5.9	5.5	7.4	0.49
MPV	3.9–11.1 fL	10.1	7.9	8.6	
PDW	%	40.5	37.2	37.9	
GRA	62–87%	86.7	91.8	89.6	
PCT	%	0.47	0.49	0.43	
HCT	37–55%	52.78	47.17	45.36	41

left atrium was also observed (Figure 4-4A). Color Doppler examination at this site showed mosaic-patterned regurgitant flow into the left atrium at end-systole (Figure 4-4B pm Video Material 4-2). The LA/Ao ratio measured at the right parasternal aortic short-axis view was 2.99:1, indicating severe left atrial enlargement as seen in ISACHC III stage patients (Figure 4-5A). In M-mode at the right parasternal left ventricular short-axis view at the mitral leaflet level, thickening of the mitral leaflet tips, a gentle slope of the F-slope, and shortening of

Table 4-2 ▸ Patient's blood biochemistry findings

Test Item	Reference Range	On Admission	One Month Later	Two Months Later	Three Months Later
GLUCOSE	60–110 mg/dL	131	128	135	154
BUN	7–25 mg/dL	28	22	15	48
CREATININE	0.3–1.4 mg/dL	1.2	0.8	1.0	1.4
ALBUMIN	2.5–4.4 g/dL	3.8	3.8	3.8	3.8
GLOBULIN	2.3–5.2 g/dL	2.7	2.9	2.8	
ALP	20–150 U/L	31	75	72	30
ALT	10–118 U/L	49	77	35	33
CALCIUM	8.6–11.8 mg/dL	11.2	11.2	10.6	10.9
PHOSPHORUS	2.9–6.6 mg/dL	2.7	4.5	4.2	2.7
AMYLASE	200–1200 U/L	551	446	518	555
K+	3.7–5.6 mg/dL	3.7	4	4.2	4.7
NA+	138–160 mg/dL	141	144	147	139
TBIL	0.1–0.6 mg/dL	0.2	0.2	0.2	0.3
TP	5.4–8.2 g/dL	6.5	6.8	6.5	6.2

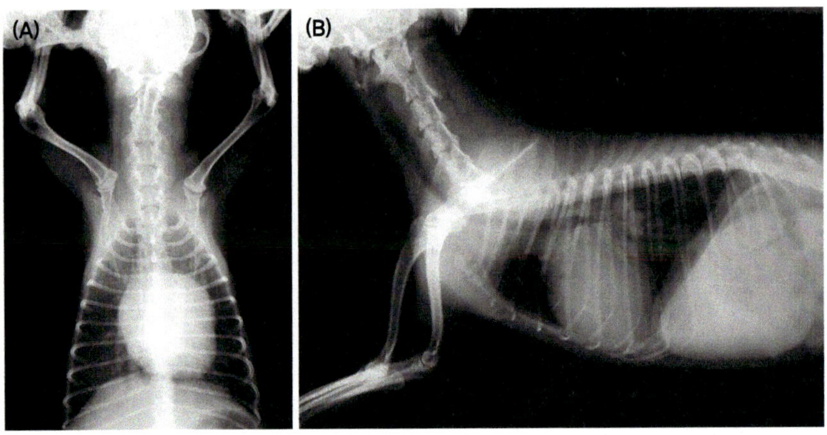

Figure 4-3 ▸ Patient's thoracic radiographs. (A) Ventrodorsal, (B) Lateral. Arrows indicate the enlarged left atrium.

the EPSS were observed (Figure 4-5B). In M-mode at the left ventricular papillary muscle short-axis view, the left ventricular internal diameter was increased, and %FS (fractional shortening) and EF (ejection fraction) appeared higher than normal due to hyper-dynamic movement caused by the volume overload and reduced afterload on the left heart (Figure 4-6A). Furthermore, the pulmonary vein flow velocity showed an increase in PVa velocity to over 0.2 m/s, suggesting left atrial volume overload (Figure 4-6B). CW Doppler measurement of mitral regurgitation velocity was 3.9 m/s (~64 mmHg PG), and dP/dT was over 1,880 mmHg, indicating left ventricular hyperkinesia (Figure 4-7A). The patient's E-peak velocity was 1.8 m/s (reference range 0.5-1.0 m/s), indicating severe left atrial pressure increase, and a ratio of 2.23:1 showed a restrictive filling pattern (Figure 4-7B). These findings seem to be due to severely enlarged left atrial dilatation. Tissue Doppler examination showed an e'/a' ratio of over 2:1, suggesting that the restrictive filling pattern observed in the transmitral flow profile was due to excessive volume overload on the left atrium rather than left ventricular hypertrophy or decreased elasticity (Figure 4-8).

Based on imaging diagnostics, this patient was finally diagnosed with **myxomatous mitral valve disease** and **heart failure classified as ISACHC IIIA or IIIB**.

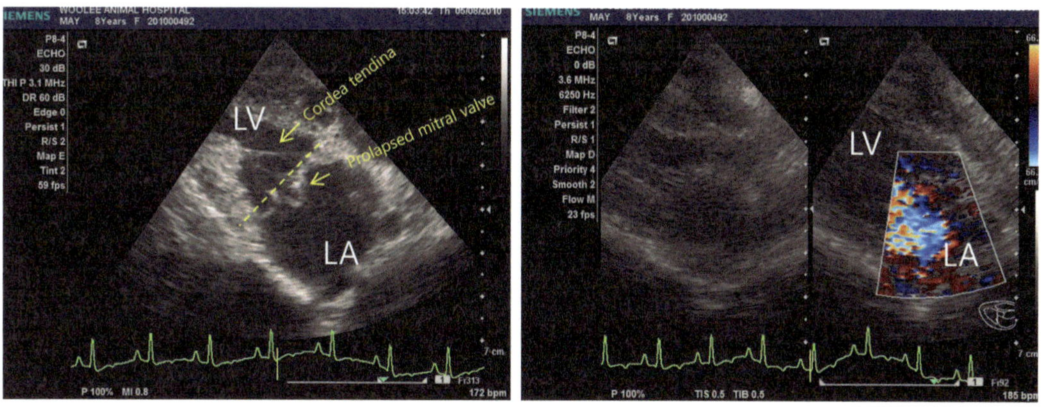

Figure 4-4 ▸ Patient's echocardiography findings. (A) B-mode left apical two-chamber long-axis view, (B) Color Doppler left apical two-chamber long-axis view.

Video Material 4-1 ▸ Mitral insufficiency. Video showing thickening and prolapse of the mitral leaflet tips.

Video Material 4-2 ▸ Mitral insufficiency. Video showing mosaic-patterned regurgitant flow into the left atrium due to mitral insufficiency.

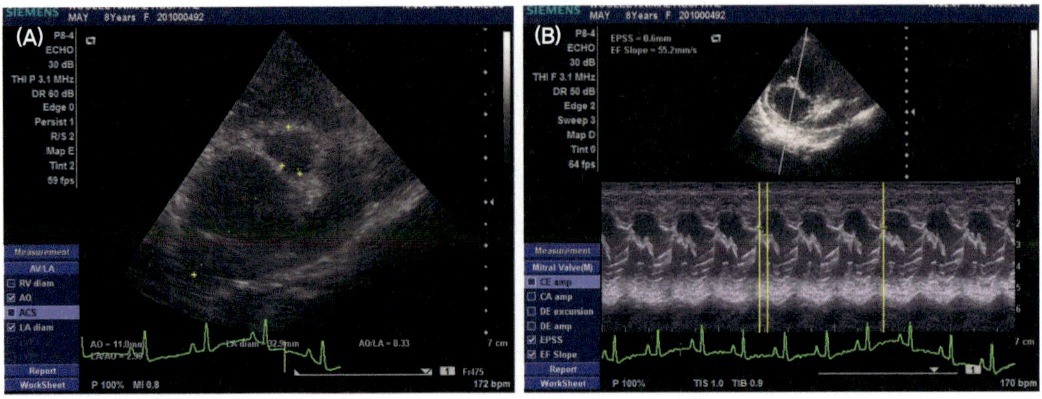

Figure 4-5 ▸ Patient's echocardiography findings. (A) B-mode right parasternal aortic short-axis view, (B) M-mode right parasternal left ventricular short-axis view at mitral leaflet level.

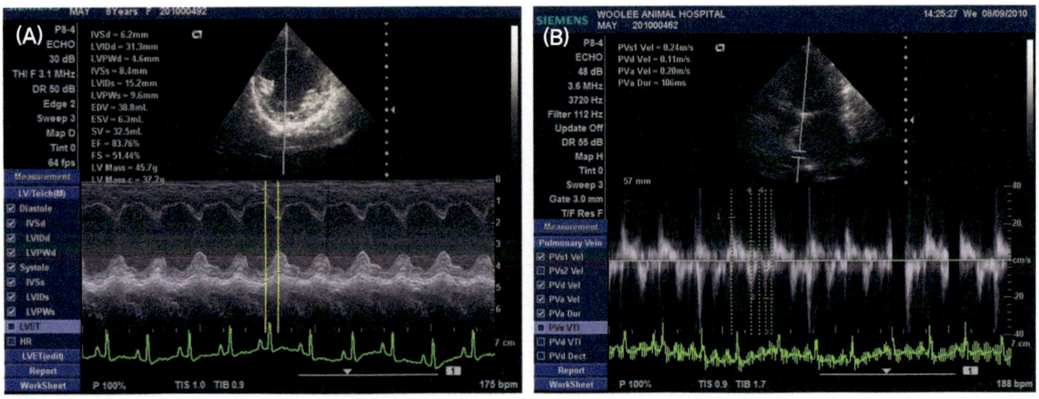

Figure 4-6 ▸ Patient's echocardiography findings. (A) M-mode right parasternal left ventricular short-axis view at papillary muscle level, (B) Pulse-wave Doppler pulmonary vein flow measurement.

Treatment and Patient Monitoring

For this patient diagnosed with mitral insufficiency, furosemide (2 mg/kg, PO, BID) was administered to reduce left atrial pressure. Enalapril (0.5 mg/kg, PO, BID) and spironolactone (1.0 mg/kg, PO, BID) were given to inhibit vascular dilation and cardiac remodeling due to RAAS activation. Pimobendan (0.3 mg/kg, PO, BID) was also administered to improve cardiac contractility and blood flow to the aorta and pulmonary artery. Since the patient's heart rate was over 180 bpm, a cough suppressant containing hydrocodone and tramadol (0.3 mg/kg, PO, BID) were prescribed instead of a bronchodilator like aminophylline. The patient's clinical symptoms significantly improved, and she no longer showed syncope.

The patient was followed up monthly. As a result, echocardiographic changes showed a gradual decrease in heart size and a gradual reduction in left heart volume overload (Table 4-3). The patient survived for over 2 years. During this period, she developed acute renal

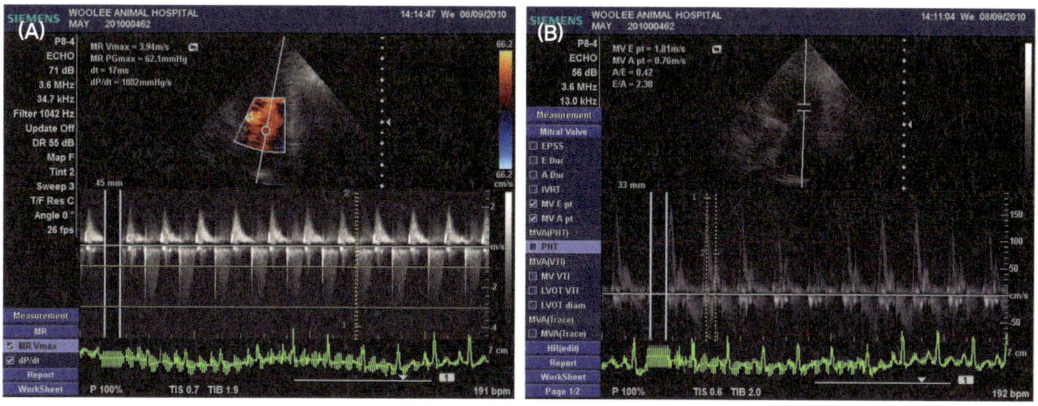

Figure 4-7 ▸ Patient's echocardiography findings. (A) CW Doppler left ventricular inflow tract flow measurement, (B) Pulse-wave Doppler left ventricular inflow tract flow measurement.

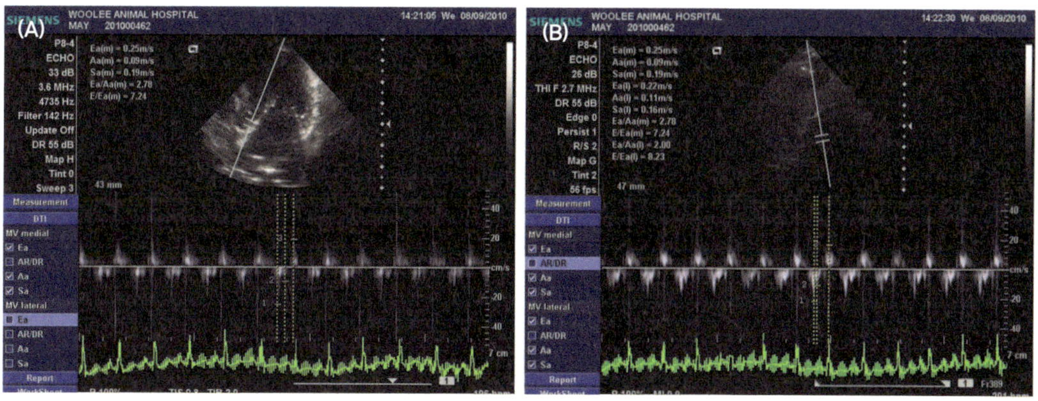

Figure 4-8 ▸ Patient's echocardiography findings. (A) Tissue Doppler at mitral annulus (medial), (B) Tissue Doppler at mitral annulus (lateral).

failure due to long-term diuretic use and recovered with peritoneal dialysis and appropriate fluid therapy at the hospital. However, she had persistent elevation of renal values due to the kidney damage incurred at that time and eventually passed away recently due to a relapse of renal failure.

Case Discussion

MMVD, a degenerative change due to aging, occurs in all breeds, but not all degenerative valve changes progress to heart failure. However, there are significant differences in the progression rate to heart failure among breeds, differences in the predilection sites for valve degeneration, and slight differences in the clinical manifestation patterns related to valve degeneration. In cases like this Maltese or Shih Tzu, the age at initial presentation is relatively younger compared to other small breeds (usually around 8-9 years old), and the disease progression is faster. On the other hand, in other small breeds (e.g., Yorkshire Terrier), the progression is slower, and valve degeneration does not readily progress to heart failure. However, in this case, it is peculiar that despite the patient's age of 9 years at presentation, she presented with severe mitral valve degeneration and severe heart failure.

In MMVD, degenerative changes occur in the mitral valve, leading to thickening and nodular degeneration of the leaflets, causing insufficiency. The regurgitation of blood into the left atrium increases the left atrial pressure and induces left atrial enlargement. This volume overload is augmented by pulmonary venous return, resulting in volume overload in the left ventricle. Ventricular volume overload causes eccentric ventricular dilation and hypertrophy. Due to the low outflow resistance into the left atrium, the left ventricle maintains a hyperdynamic contraction pattern. However, this chronic volume load and maladaptive neurohormonal system activation (including RAAS) promote progressive remodeling. The resulting increased circulating blood volume and venous pressure elevations eventually lead to congestive heart failure (CHF). As a result, when the hydrostatic pressure in the pulmonary vasculature increases, pulmonary edema, characteristic of congestive left heart failure, occurs with fluid retention in the interstitium.

This patient presented with coughing and intermittent syncope. On physical examination, a systolic regurgitant murmur was confirmed at the left apical region. Considering these symptoms and age, mitral insufficiency was suspected. ECG findings, such as P mitrale, provided indirect evidence of left atrial enlargement. Radiographic findings of dorsal tracheal displacement, cardiomegaly, left atrial enlargement, and pulmonary vein distention were commonly seen in patients with mitral insufficiency. On echocardiography, mitral valve prolapse and nodular degeneration were observed, along with severe regurgitation into left atrium with an MR fraction of over 70%, severe left atrial enlargement with an LA:Ao ratio of almost 3:1, and severe left ventricular enlargement with an E-peak velocity approaching 1.8 m/s. Additionally, an LVIDd/Ao ratio of 2.84:1 indicated severe left ventricular enlargement. Based on a comprehensive review of these test results, the patient was diagnosed with degenerative mitral insufficiency and heart failure of ISACHC IIIA-IIIB grade.

There are several diagnostic criteria for MMVD on echocardiography. Representative criteria include: 1) Mitral regurgitation (MR) fraction, which measures the degree of blood regurgitation from the left ventricle to the left atrium using color Doppler; 2) LA:Ao ratio, which evaluates the degree of left atrial enlargement by measuring the ratio of the left atrium to the aorta; 3) E-peak velocity (E wave: ventricular diastolic filling wave), which measures the velocity of blood inflow from the LA to the LV and is a crucial indicator of increased left atrial pressure when high; 4) Mitral regurgitation (MR) velocity, which measures the velocity of blood regurgitation from the left ventricle to the left atrium; and 5) Left atrial (LA) diameter measurement method. However, the fourth method is variable depending on the patient's condition or excitement, and the fifth method is unreliable for evaluating the degree of left atrial enlargement due to a large margin of error in the left atrial size reference. For other echocardiographic findings that evaluate prognosis, please refer to the main text.

At our hospital, there are several criteria for determining the type and dosage of med-

Table 4-3 ▸ Changes in patient's echocardiographic measurements

	Day 0	Day 30	Day 60	Day 90		Day 0	Day 30	Day 60	Day 90
EPSS	0.6 mm	0.6	0.8	0.0	IVSd	6.2 mm	5.6	5.7	5.0
EF Slope	55.2 mm/s	101.2	54.5	72.3	LVIDd	31.3 mm	27.3	29.7	29.4
LA/Ao ratio	2.99	2.46	2.52	2.33	LVPWd	4.6 mm	6.5	5.3	5.0
MR Vmax	4.52 m/s	3.94	4.87	4.62	IVSs	8.4 mm	10.8	6.4	7.8
MR PGmax	81.7 mmHg	62.1	66.3	85.66	LVIDs	15.2 mm	14.9	16.3	14.4
dt	28 ms	17	44		LVPWs	9.6 mm	10.5	8.8	9.4
dP/dt	1142 mmHg/s	1880	727		EDV	38.8 mL	27.8	34.2	
E peak	1.74 m/s	1.81	1.37	1.49	ESV	6.3 mL	6.0	7.5	
A peak	0.78 m/s	0.76	0.47	0.88	SV	32.5 mL	21.8	26.7	27.87
E/A	2.23	2.38	2.91	1.70	EF	83.76%	78.42	78.07	83.66
Ea(m)	0.23 m/s	0.25		1.07	FS	51.44%	45.42	45.12	51.02
Aa(m)	0.13 m/s	0.09		0.84	LV mass	45.7 g	42.4 g	42.9	30.34
Sa(m)	0.29 m/s	0.19		0.14	LV mass-c	37.2	34.6	42.9	30.34
Ea(l)		0.22 m/s		1.6	AV Vmax	0.86 m/s			
Aa(l)		0.11 m/s		0.76	AV PGmax	3.0 mmHg			
Sa(l)		0.16 m/s		1.4	PV regurg				1.17 m/s
PVs		0.24 m/s			LVIDd /Ao	2.84	2.48	2.7	2.69
PVd		0.11 m/s							
PVa		0.20 m/s							

ication when a MMVD patient first presents. If the patient shows left or right atrial enlargement on radiographs, diuretics are prescribed. All patients receiving diuretics are also prescribed ACE inhibitors to block RAAS activation. Pimobendan is also administered to asymptomatic patients with cardiomegaly. Spironolactone is additionally prescribed only for patients with severe ascites or fluid retention. The initial dose of furosemide is based on the LA:Ao ratio and E wave peak velocity; mainly, patients with an LA: Ao ratio of 2:1 or higher, and E wave peak velocity greater than 1.2 m/s receive 2 mg/kg BID or more, while patients with a lower ratio receive 1.0-1.5 mg/kg BID, depending on the LA:Ao ratio and the patient's symptoms. For patients with concomitant chronic kidney disease or a high risk of developing it, BID-TID-QID administration is recommended rather than BID. This is to reduce kidney damage by reducing the single dose of diuretic and shortening the dosing interval. Pimobendan is usually indicated when the LVIDd/Ao ratio is 2:1 or higher, and the initial dose is started at 0.25-0.3 mg/kg BID, increasing to 0.5-0.75 mg/kg BID or 0.5 mg/kg TID if the patient's condition worsens. In this patient's case, the LA:Ao ratio was ~3:1, the E peak velocity was 1.8 m/s, and the LA:LVIDd was ~2.9:1 at presentation, indicating severe heart failure. Typically, such patients would be prescribed furosemide 3 mg/kg BID and pimobendan 0.5 mg/kg BID. However, since the patient's vitals were relatively stable and no pulmonary infiltrates (pulmonary edema) were observed, furosemide 2 mg/kg BID and pimobendan 0.3 mg/kg BID were prescribed for 2 weeks. Fortunately, a phone check after 2 weeks revealed that the patient's cough had significantly decreased, she no longer showed syncope, and her vitality was excellent. Therefore, the same level of medication was instructed to be continued, and monthly regular check-ups were scheduled. Fortunately, the patient's echocardiogram one month later showed a favorable decrease of about 20% in LA:Ao ratio, LA:LVIDd, and E-peak, suggesting that the initial dosage was appropriate. No significant changes were observed in the two-month check-up, so the dosage was maintained. However, a borderline increase in renal values was observed at the three-month check-up, and to reduce this, increasing water intake and administering renal supportive agents like Azodyl® and Rubenal® were recommended. Two years later, the patient presented with acute on chronic kidney disease due to side effects of ear medication (neomycin and gentamicin antibiotic treatment) and recovered with dialysis and careful fluid therapy. It is presumed that neomycin, commonly found in general otitis externa medications, caused severe nephrotoxicity in a patient taking diuretics.

The general survival period for MMVD patients is more than 3 years for ISACHC I heart failure, 1-3 years for ISACHC II heart failure, ~1 year for ISACHC IIIa, and less than 6 months for ISACHC IIIb heart failure. This prognosis can vary greatly depending on the type of medication, monitoring schedule, owner's diligence, and the patient's will to live. At our hospital, the patient's prognosis is primarily evaluated based on echocardiographic findings. Patients with findings such as LA:Ao ratio >2.0, LVIDd:Ao ratio >2.5, E-peak velocity >1.2 m/s, and MR fraction >75% are classified as ISACHC III or higher stage and receive intensive treatment and management. Owners are informed of an expected lifespan of less than one year. Of course, many such patients survive for over 2 years.

CASE 2

Patient Information

The patient is a 14-year-old male Maltese referred to Kangwon National University Cardiac Center for coughing and trembling. He had been treated for a cold at a local hospital for coughing as a main symptom and later received emergency treatment for dyspnea.

Physical Examination and Phonocardiogram

On auscultation, a grade V-VI/VI systolic regurgitant murmur was heard at the left apical region, with a precordial thrill palpable on the thorax. Systolic blood pressure was 115 mmHg, heart rate was 145-150 bpm, respiratory rate was 66 bpm, and body temperature was 37.5 degrees Celsius. The patient's phonocardiogram showed a holosystolic plateau-shaped systolic murmur at the left apex (Figure 4-9).

Diagnostic Tests

Electrocardiogram (ECG) Examination: Heart rate was 145-150 bpm, rhythm origin was normal sinus rhythm (NSR). The P-wave morphology was wide and notched, indicat-

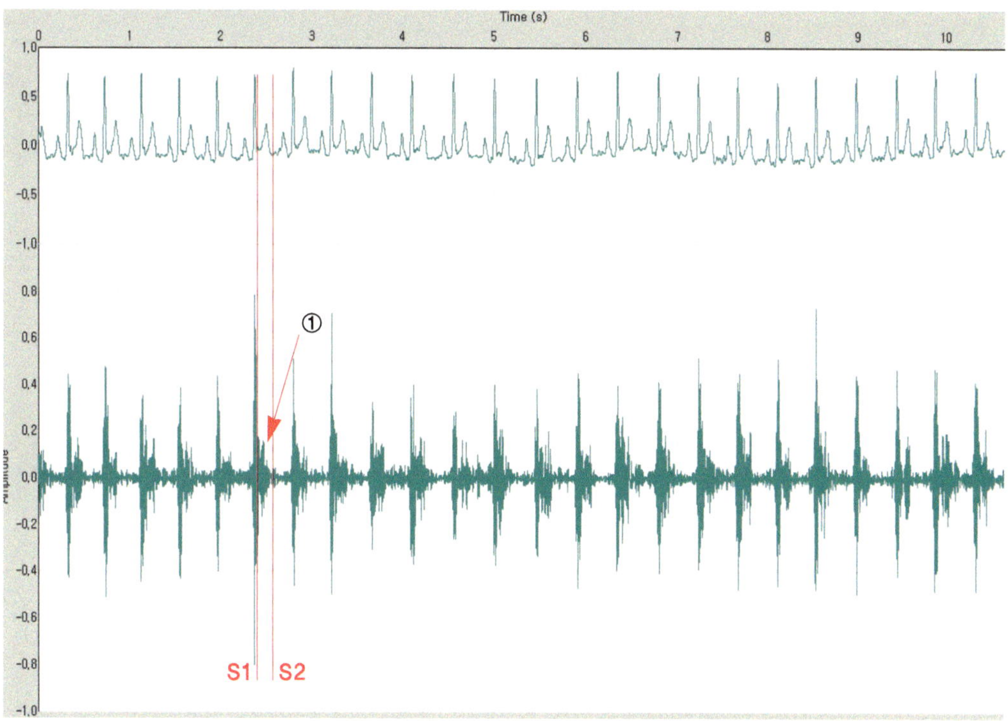

Figure 4-9 ▸ Patient's phonocardiogram (recorded from the left apical region). The patient showed a typical systolic decrescendo regurgitant murmur (arrow ①).

ing left atrial enlargement, and the R-wave morphology showed a tall R wave (>2.5 mV), indicating left ventricular hypertrophy. The patient's R-R interval was regularly regular, and the PR interval was constant (Figure 4-10).

Laboratory Tests: Complete blood count and biochemistry revealed no abnormalities other than hypoalbuminemia, azotemia, and elevated liver enzymes (Table 4-4). However,

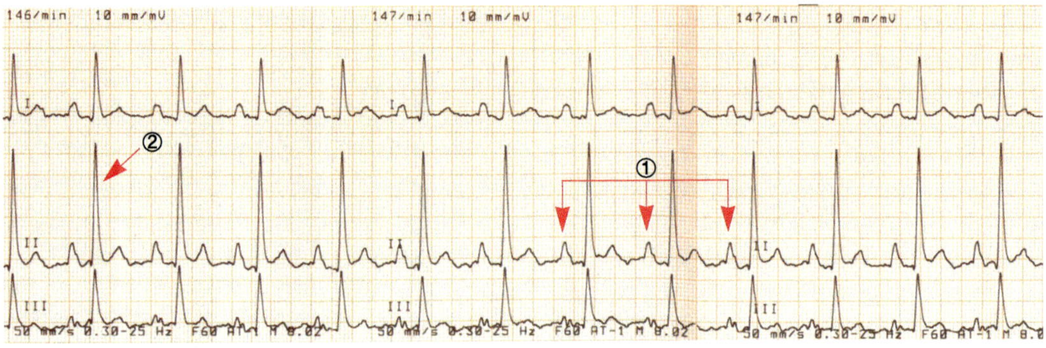

Figure 4-10 ▸ Patient's electrocardiogram findings. Shows typical left atrial enlargement with a wide and notched P wave (arrow ①), and tall R wave (arrow ②) indicating left ventricular hypertrophy.

Table 4-4 ▸ Patient's blood test findings

Test Item (CBC)	Reference Range	Result	Test Item (Chemistry)	Reference Range	Result
WBC	6–17 K/uL	9.52	TP	5.6–7.1 g/dL	6.0
NE	3–11 K/uL	9.08	ALB	3.1–4.1 g/dL	2.7
LY	1–5 K/uL	0.32	ALP	12–122 U/L	135
MO	0.1–2 K/uL	0.09	AST	16–50 U/L	15
EO	0–1.3 K/uL	0.02	ALT	3–100 U/L	20
BA	0–0.1 K/uL	0.01	Cre	0.5–1.3 mg/dL	2.9
RBC	5.5–8.5 M/uL	6.90	BUN	8–25 mg/dL	>200
Hb	12–18 g/dL	15.1	GLU	60–120 mg/dL	119
PCV	37–55 %	40.8	Ca	7.2–12.8 mg/dL	10.5
MCV	60–77 fL	59.2			
MCHC	33–36 g/dL	37			
MCH		21.9			
PLT	150–500 K/uL	292			

the kidney function indicated IRIS Stage III chronic kidney disease (IRIS III: creatinine 2.9-5.0 mg/dL ± revised IRIS II).

Thoracic Radiography: Thoracic radiographs showed cardiomegaly (VHS = 10.5 V, L = 6.0 V, S = 4.5 V), left atrial enlargement, and left heart system enlargement due to dorsal tracheal displacement (Figure 4-11). The patient's thoracic radiographs were similar to typical findings in mitral insufficiency.

Echocardiographic Examination: On 2D echocardiography, the mitral valve appeared abnormally thickened, mitral valve prolapse was observed in the apical view (Video Material 4-3), and abnormal mitral valve closure caused mitral insufficiency. Additionally, the left atrium and left ventricle appeared significantly enlarged (Figure 4-12). On color Doppler echocardiography, mosaic-patterned turbulent flow was observed throughout the left atrium (Video Material 4-4). This indicates very severe MMVD (Figure 4-13). On CW Doppler

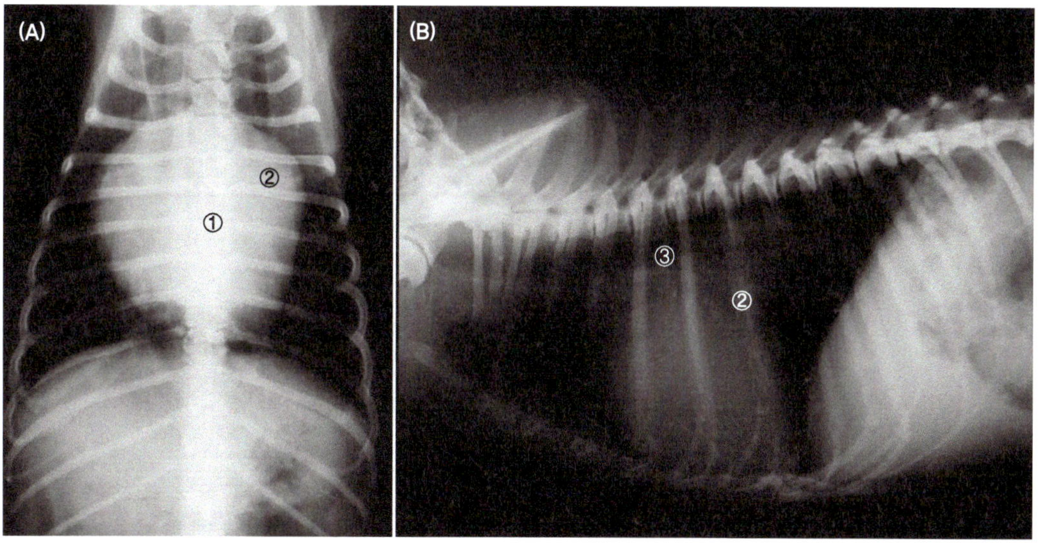

Figure 4-11 ▸ Patient's radiographic findings. The DV thoracic radiograph showed overall cardiomegaly (A-①) and VHS was measured at 10.5V (L = 6.0V S = 4.5V). Left atrial enlargement (A-②) was also observed. Lateral radiographs showed left atrial enlargement (B-②) and dorsal displacement of the trachea (B-③).

Video Material 4-3 ▸ Cardio-renal syndrome. Valve degeneration, which is the cause of mitral insufficiency, can be confirmed.

Video Material 4-4 ▸ Cardio-renal syndrome. Mosaic blood flow regurgitating into the left atrium due to mitral insufficiency can be confirmed.

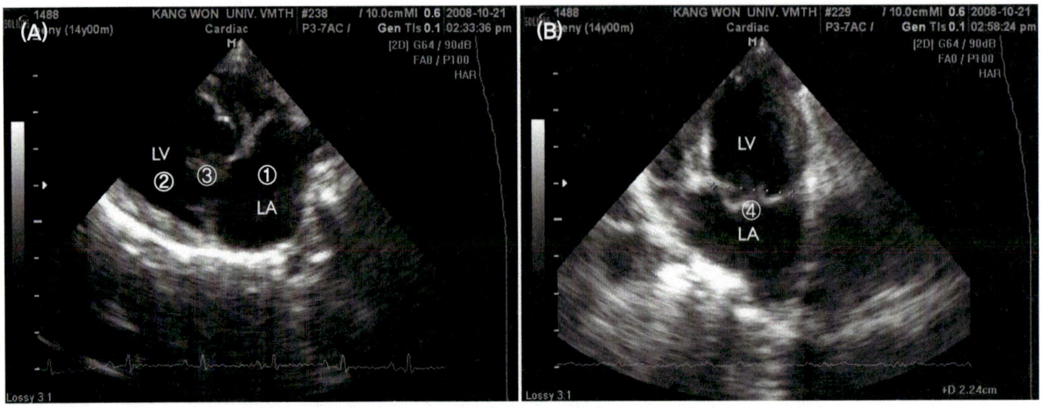

Figure 4-12 ▸ Patient's 2D ultrasound 4-chamber long-axis view (A) and apical view (B). Left atrial enlargement (A-①) and left ventricular enlargement (A-②) are observed. The mitral valve between the left atrium and left ventricle is markedly hypertrophied (A-③). Additionally, mitral valve prolapse into the left atrium is observed (B-④).

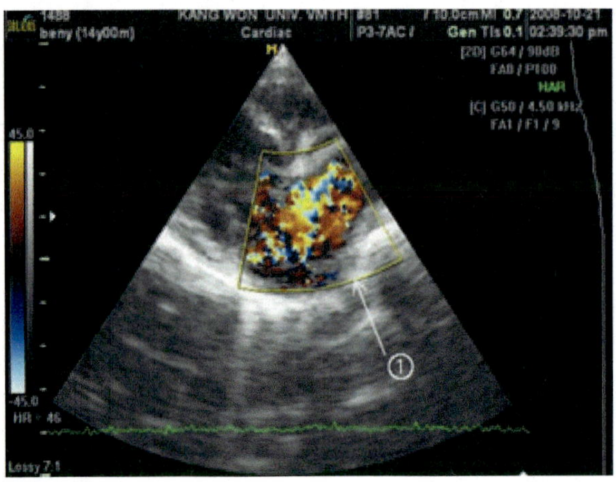

Figure 4-13 ▸ Patient's 4-chamber long-axis color Doppler findings. Very severe mosaic-patterned regurgitation (arrow ①) occupying more than 2/3 of the left atrium is observed.

echocardiography, severe regurgitation into the left atrium was confirmed through Doppler examination from the left ventricle to the left atrium, and the E-peak velocity of blood flow from the left atrium to the left ventricle was measured at 2.4 m/s (reference range 0.8-1.0 m/s). The prognosis for patients with such high E-peak velocity is generally poor (Figure 4-14).

The patient was found to have azotemia, so renal ultrasound was performed. The hyperechoic changes in the renal cortex, including the medullary rim sign, may suggest acute kidney injury (Video Material 4-5). Generally, such changes occur due to ethylene glycol toxicity or melanin toxicity, however, it may not be specific to renal dysfunction. In this patient, acute renal failure secondary to heart failure was suspected. Therefore, correction of renal failure was also necessary simultaneously with heart failure treatment (Figure 4-15).

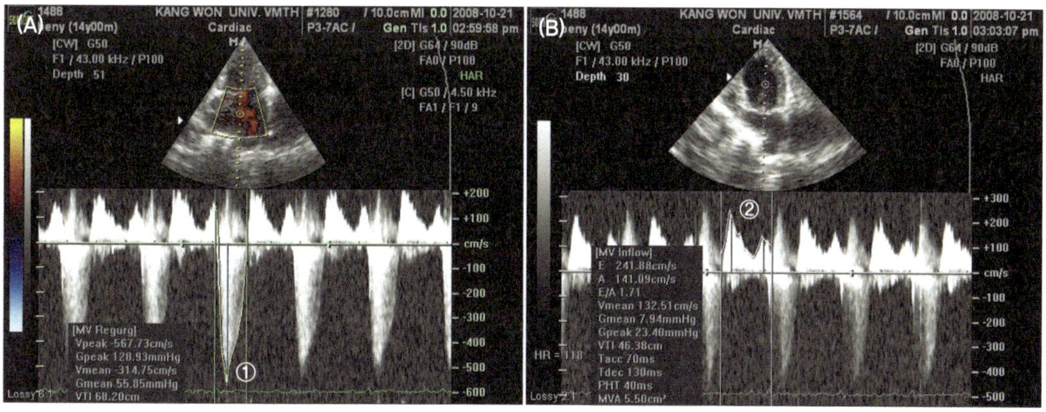

Figure 4-14 ▸ Patient's CW Doppler findings. Regurgitation from the left ventricle to the left atrium (5.6 m/s) was observed (A-①). Doppler measurement of MV inflow from the left atrium to the left ventricle showed E-peak velocity of 2.4 m/s (B-②).

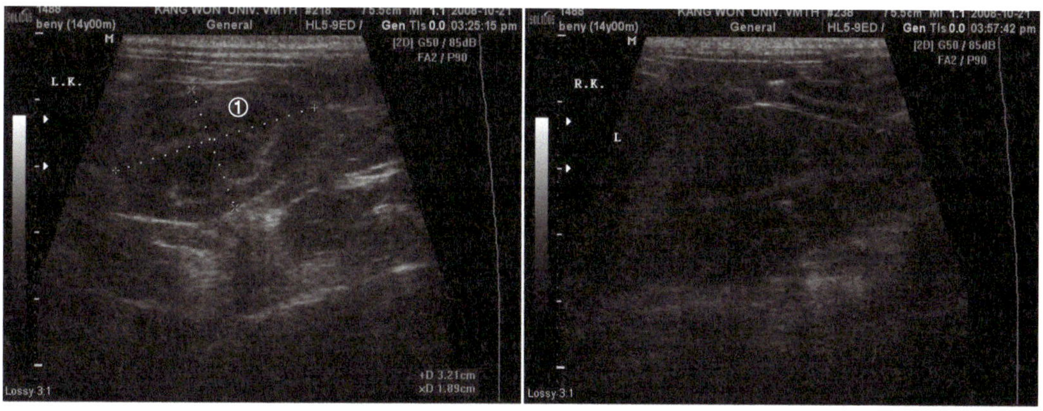

Figure 4-15 ▸ Patient's renal ultrasound findings. A hyperechoic image (①) was confirmed at the corticomedullary junction in the left kidney. The hyperechoic changes in the renal cortex, including the medullary rim sign, may suggest acute kidney injury.

Video Material 4-5 ▸ Cardio-renal syndrome. Renal ultrasound findings showing a halo sign formed between the cortex and medulla.

On physical examination, the patient showed symptoms of coughing and shivering, and auscultation revealed a left apical regurgitant murmur, suggesting mitral insufficiency. ECG findings showed a wide and notched P-wave, typical for left atrial enlargement, and a tall R-wave, indicating left ventricular hypertrophy. Radiographs showed overall cardiomegaly, left atrial enlargement, and left heart system enlargement due to dorsal tracheal displacement. Echocardiography revealed thickening and prolapse of the mitral valve leaflets. Color and CW Doppler examinations showed severe regurgitation with a mosaic pattern occupying more than 2/3 of the left atrium. The E-peak velocity into the left ventricle was 2.4 m/s, which was faster than normal, and regurgitant flow from the left ventricle to the left atrium was very fast at 5.6 m/s. Additionally, laboratory tests revealed azotemia, and renal ultrasound showed a hyperechoic image (medullary rim sign) at the corticomedullary junction of the left kidney, suspected of being due to antifreeze (ethylene glycol) toxicity or melanin toxicity. Based on a comprehensive review of the above findings, the patient was finally diagnosed with mitral insufficiency complicated by renal failure.

Treatment and Patient Monitoring

To correct the azotemia in this patient, peritoneal dialysis was performed (Figure 4-16 pm Video Material 4-6). After a total of 13 peritoneal dialyses over 3 days, the patient's azotemia improved to some extent with creatinine 2.2 mg/dL and BUN 126 mg/dL (Table 4-5). Subsequently, for mitral insufficiency, furosemide (2.5 mg/kg, PO, BID) was administered to reduce left atrial pressure. Enalapril (0.5 mg/kg, PO, BID) and spironolactone (1 mg/kg, PO, BID) were given to inhibit vascular dilation and cardiac remodeling due to RAAS activation and to prevent excessive potassium excretion. Pimobendan (0.3 mg/kg, PO, BID) was administered to improve left heart contractility and blood flow to the aorta. As a measure for renal failure, azodyl (1 capsule/dog, PO, BID) and kremezine (200 mg/dog, PO, BID) were additionally prescribed. With these drug therapies, the patient's condition further improved.

After approximately 3 months of drug treatment, evaluation of the patient showed an increase in body weight to 4.6 kg, and azotemia significantly improved (creatinine 1.1 mg/dL, BUN 55 mg/dL). On echocardiography, the patient's condition significantly improved.

Case Discussion

Diagnostic features of mitral insufficiency include cardiomegaly with left atrial enlargement on thoracic radiographs, systolic regurgitant murmur at the left apex, a wide and notched P-wave on ECG, left atrial and left ventricular enlargement on echocardiography, and turbulent flow regurgitating into the left atrium during systole on color echocardiography. This case is a typical example of mitral insufficiency demonstrating all these diagnostic features. Additionally, this patient also had renal failure with severe azotemia confirmed by laboratory tests, along with hyperechoic lesions at the corticomedullary junction of the kidney. This case presented with severe mitral regurgitation and kidney injury due to heart failure, so correction of azotemia through peritoneal dialysis was prioritized. After 13 peritoneal dialyses over 3 days to stabilize azotemia and the patient's condition, drug treatment for

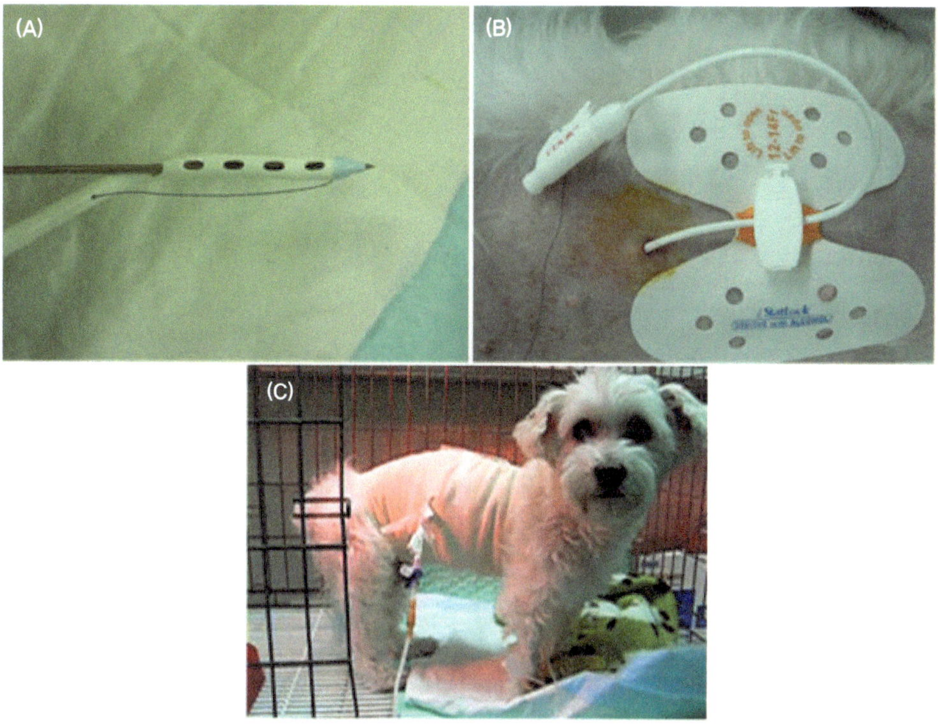

Figure 4-16 ▶ Peritoneal dialysis set and peritoneal dialysis setup. A peritoneal dialysis set was placed in the abdominal cavity via the abdominal wall to improve the patient's azotemia. (A) Silicone peritoneal dialysis catheter, (B) Butterfly bandage for securing the dialysis catheter to the abdominal wall, (C) Peritoneal dialysis in progress.

Video Material 4-6 ▶ Video showing the peritoneal dialysis process.

mitral insufficiency causing left heart failure, and azodyl and kremezine were continuously prescribed to alleviate renal failure. After 3 months, the patient's condition was confirmed to have improved to some extent through laboratory tests and echocardiography.

While surgical treatment involving valve replacement is an option for mitral insufficiency patients, surgical procedures are not yet common in animals. Therefore, most cases of mitral insufficiency are managed by medical treatment. The goal of pharmacotherapy is to improve the patient's quality of life. Diuretics and angiotensin-converting enzyme inhibitors (ACEi) are the main drugs used for this medical management. However, high-dose diuretic

Table 4-5 ▶ Changes in blood test results during peritoneal dialysis

Chemistry Test Item	Reference Range	Before Dialysis	After 3 Dialyses	After 8 Dialyses	After 13 Dialyses
TP	5.6−7.1 g/dL	6.0	6.3	6.0	5.4
ALB	3.1−4.1 g/dL	2.7	3.0	2.6	2.4
ALP	12−122 U/L	135	143	144	126
AST	16−50 U/L	15	18	19	23
ALT	3−100 U/L	20	22	13	14
Cre	0.5−1.3 mg/dL	2.9	2.7	2.6	2.2
BUN	8−25 mg/dL	>200	170	154	126
GLU	60−120 mg/dL	119	190	112	110
Ca	7.2−12.8 mg/dL	10.5			

therapy to completely block fluid retention is not desirable for long-term patient management. Therefore, inodilators like pimobendan, as in this case, can be very useful for treating mitral insufficiency. Especially in patients with azotemia, high-dose diuretics can worsen drug-induced kidney injury, so the application of pimobendan is a good way to avoid high-dose diuretic therapy. Thus, as seen in this case, inodilators like pimobendan can be useful in treating left heart failure caused by mitral insufficiency.

The prognosis of mitral insufficiency is determined by the degree of valve abnormality and cardiac remodeling. In this patient's case, severe left atrial enlargement and severe left heart failure symptoms equivalent to ISACHC IIIa were observed. Therefore, if left untreated, the patient would likely have a survival probability of less than 6 months. In this case, the patient's condition significantly improved with consistent pharmacotherapy.

A retrospective study in human medicine found that 60% of 755 patients with chronic heart failure also had chronic kidney disease, and a close relationship between glomerular filtration rate and the prognosis of chronic heart failure has been reported. Recent veterinary reports also indicate that the more severe the heart failure, the more frequently azotemia and kidney damage occur. Since the kidneys receive 25% of the blood ejected from the heart, the two organs share various homeostatic mechanisms hemodynamically and neuro-hormonally. The problem is that while the homeostatic responses of the two organs do not affect each other in normal cases, if there is renal failure or heart failure, these homeostatic responses can adversely affect the other organ. For example, in congestive heart failure, diuretic prescriptions to reduce left atrial volume overload greatly help alleviate heart disease symptoms, but they reduce renal blood flow by decreasing circulating blood volume, thereby decreasing glomerular filtration rate, and if the patient has renal failure, renal perfusion can decrease, worsening renal failure. Conversely, patients with severe renal failure often have high intraglomerular pressure and glomerular hypertension due to glomerular fibrosis, etc. Ultimately, this glomerular hypertension often leads to systemic hypertension. In a state of systemic hypertension, the left ventricle experiences increased pressure load, increasing the workload on the left ventricular wall and increasing wall stress. If not properly treated, the left ventricular wall undergoes concentric hypertrophy (a compensatory response of the heart to hypertension, characterized by markedly thickened left ventricular walls without enlargement of the ventricular lumen) and eventually leads to heart failure. However, in severe MR dogs, the increased systemic blood pressure may not cause concentric hypertrophy because the increased LV systolic pressure (due to increased arterial pressure) would be released through severe MR.

In heart failure patients, an increase in blood creatinine levels is often observed with long-term use of ACE inhibitors like enalapril. This is because ACE inhibitors dilate renal efferent arterioles and reduce intraglomerular pressure, putting the kidneys in a hypoperfused state. This eventually reduces the glomerular filtration rate and can increase blood creatinine levels. However, prospective and retrospective studies that applied various ACE inhibitors to heart failure patients show that the positive effects of ACE inhibitors on cardiac and renal function are greater. ACE inhibitors slow the progression of renal parenchymal fibrosis and have a vasodilatory effect, reducing damage to glomerular vessels and alleviating proteinuria. The negative effects of ACE inhibitors on the kidneys are also temporary, and in a study that tested the efficacy of enalapril long-term, the incidence of azotemia in heart failure patients treated with the regular dose of enalapril (0.5 mg/kg) was lower than in the control group.

The treatment and management of patients with cardio-renal syndrome are very challenging, and effective treatment methods are not yet properly established. Ultimately, regular monitoring of renal function is important under the assumption that all patients receiving long-term heart failure treatment will develop renal dysfunction.

Since the most effective treatment for cardio-renal syndrome patients is prevention, it is crucial to establish an appropriate heart failure treatment strategy tailored to the patient's heart failure progression stage and, above all, to optimize the diuretic dosage. Such strategies include administering small divided doses rather than a large single dose (BID-TID), combining with or switching to a different diuretic if resistance to one diuretic is observed, and adding other heart failure medications (e.g., spironolactone, nitrate patch, amlodipine) to reduce diuretic usage. Most importantly, it is crucial to thoroughly monitor the patient to observe the response to diuretic dosage and use the optimal diuretic dose for the patient.

As part of periodic renal monitoring, measuring the renal resistive index (RI value) via ultrasound can, to some extent, evaluate whether renal blood flow resistance is increasing. However, once renal failure has occurred and the patient presents, it is often difficult to apply the usual renal failure treatment methods (aggressive fluid therapy) as they are. Aggressive fluid therapy in volume-overloaded patients can worsen congestive heart failure and induce pulmonary edema. Therefore, in such cases, lower doses of fluid therapy than in typical renal failure patients (judiciously administering only maintenance fluid, 30 mL × body weight (kg) + 70 mL at a rate of 5-10 mL/kg/hour) or performing peritoneal dialysis can be effective. However, what is most important is to make the owner aware that patients with heart failure have a high risk of developing renal failure, so periodic monitoring is essential. Regular monitoring of blood tests, urine tests, UPC tests, and blood pressure should be carefully performed to prevent the patient from presenting in a uremic state.

Generally, the prognosis varies greatly depending on the severity of heart failure. Most patients have received long-term diuretic prescriptions previously, so the long-term prognosis is generally not good.

CASE 3

Patient Information

The patient is a 12-year-old female Maltese who presented to a local hospital with sudden coughing, as if she had choked, and was diagnosed with cardiogenic pulmonary edema and received emergency treatment. She was subsequently diagnosed with mitral valve chordae tendineae rupture on ultrasound and was referred for accurate diagnosis and treatment, being told she would only survive for 2-3 months.

Physical Examination

On physical examination, a systolic regurgitant murmur (4/6) was auscultated at the left apex, and systolic blood pressure was 120 mmHg. No other abnormal findings were observed on physical examination.

Diagnostic Tests

Laboratory Evaluation: Blood tests showed a mild decrease in PCV (34.9%, RR 37-55), mild azotemia (BUN 55 mg/dL RR 7-27, creatinine 1.5 mg/dL, RR 0.5-1.8), and mild elevation of liver enzymes (ALT 335 U/L, RR 10-100). Biomarker tests revealed NT-proBNP 5105 (RR 0-900), which may suggest congestive heart failure, and SDMA 31 (RR 0-14), indicating IRIS Grade I AKI. Spec cPL was 38 (RR 0-200).

Electrocardiogram (ECG) Examination: Left ventricular enlargement pattern was observed, with no other specific findings.

Thoracic Radiography: Thoracic radiographs showed cardiomegaly (VHS 11.5, VLAS 3, RLAS 2.6) and severe left atrial enlargement, but no bronchial compression or pulmonary infiltrates. However, dorsal tracheal elevation was observed (Figure 4-17).

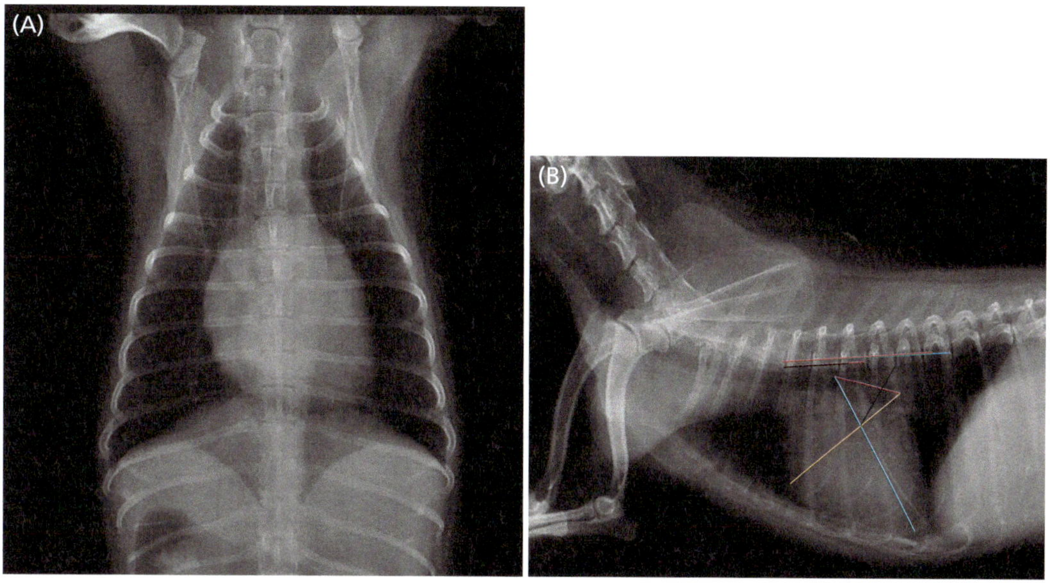

Figure 4-17 ▶ Thoracic radiographs (A: Ventrodorsal, B: Right lateral). Blue (VHS long axis), orange (VHS short axis), red (VLAS), black (RLAS).

Echocardiographic Examination: LA/Ao SX was measured at 2.5 (ISACHC III), LVIDdN at 1.57 (normal range), and LVIDd/Ao at 2.17 (ISACHC III). Color Doppler examination showed mosaic regurgitant flow into the left atrium, and 2D echocardiography confirmed rupture of the chordae tendineae supporting the A1 leaflet of the mitral valve (Figure 4-18). PW Doppler examination showed mitral flow E-peak velocity of 0.88 m/sec (ISACHC I), and CW Doppler examination confirmed high-velocity mitral regurgitation flow. TDI Doppler examination revealed septal E/E' 16.67 and lateral E/E' 10.0, with an average E/E' of 13, indicating congestive heart failure (Figure 4-19). M-mode aortic short-axis measurement of %LA-FS was 27%, indicating heart failure beyond ACVIM Stage C (Figure 4-20). PW Doppler examination of the right ventricular outflow tract flow showed AT 53 m and AT/ET 0.53, with no findings suggesting pulmonary hypertension (Figure 4-20). M-mode aortic short-axis measurement of %RV-FS was also 77%, with no findings suggesting pulmonary hypertension (Figure 4-20). PW Doppler examination at the left apical region showed IVRT 35 ms, LVET 163 ms, E/IVRT 2.51, and LV Tei index of 0.45, indicating impairment of both ventricular diastolic and systolic function and suggesting congestive heart failure (Figure 4-20).

Treatment and Patient Monitoring

Based on thoracic radiographs showing cardiomegaly and left atrial enlargement, echocardiography showing atrial and ventricular enlargement, findings suggesting increased left atrial pressure, and mitral regurgitation, the patient was diagnosed with mitral valve insufficiency accompanied by ACVIM C (ISACHC IIIa) stage congestive heart failure. Additionally, based on SDMA levels, she was diagnosed with concomitant IRIS Stage II chronic kidney

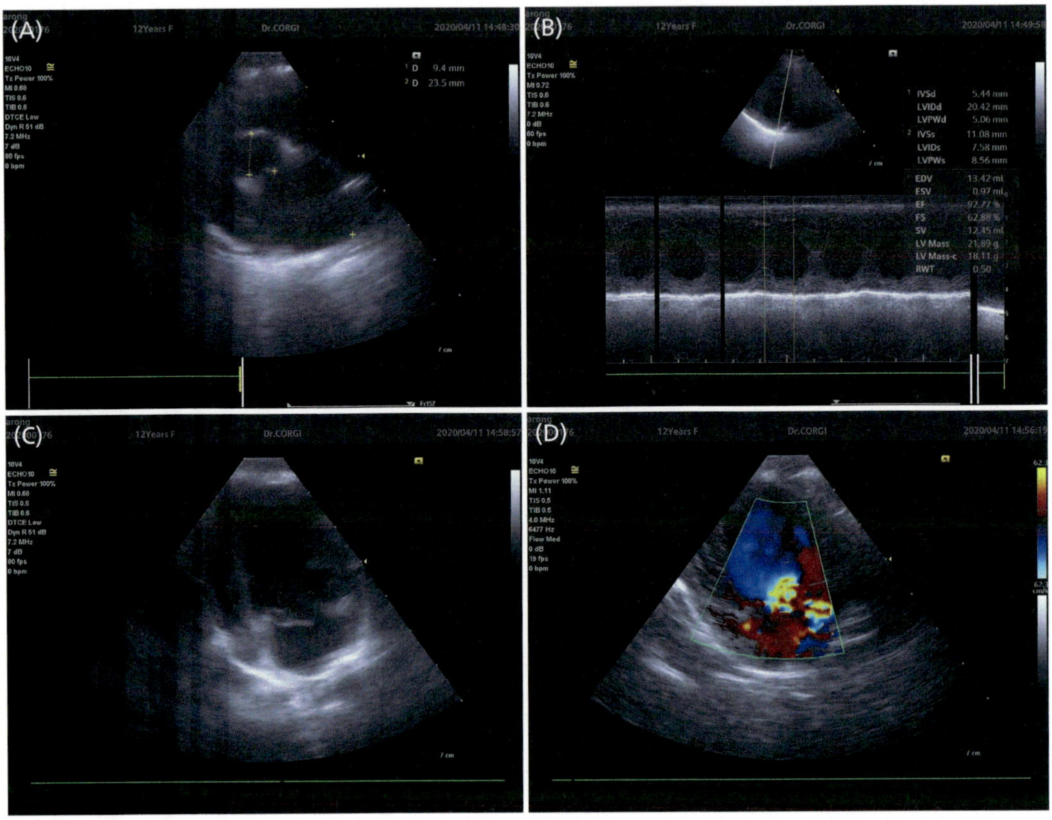

Figure 4-18 ▸ Echocardiography. A: LA/Ao was measured at 2.5 on the aortic short-axis view. B: M-mode left ventricular papillary muscle short-axis view showed LVIDd of 20.42 mm, LVIDdN of 1.57, and LVIDd/Ao of 2.17. C: Rupture of the chordae tendineae supporting the A1 leaflet of the mitral valve is observed in the left apical long-axis view. D: Color Doppler examination shows mosaic regurgitant flow into the left atrium.

disease. The patient's left ventricular enlargement was relatively mild compared to left atrial enlargement, and acute pulmonary edema was presumed to have occurred due to rupture of the chordae tendineae supporting the mitral valve. Since the patient had been taking high-dose diuretics for a long time at a local hospital, IRIS Grade I AKI was presumed to have occurred. Therefore, treatment focused on minimizing diuretic dosage. Initially, the patient was prescribed torsemide (0.15 mg/kg, q12h), enalapril (0.5 mg/kg, q12h), spironolactone (1 mg/kg, q12h), and pimobendan (0.3 mg/kg, q12h). Electrolyte and renal function tests were scheduled for 2 weeks later to monitor for hypokalemia or hyponatremia due to torsemide, and daily sleeping respiratory rates were recorded to check for recurrence of pulmonary edema. Thoracic radiography at a local hospital was instructed 2 weeks after medication. Two weeks later, thoracic radiographs showed no change in cardiomegaly but no pulmonary infiltrates, and the patient's sleeping respiratory rate remained between 17-25 breaths/minute. Electrolyte tests showed normal potassium and sodium levels, but chloride was decreased (99 mmol/L, RR 102-117), and phosphorus (6.7 mg/dL, RR 1.9-5) and calcium (12.6 mg/dL, RR 9.3-12.1) levels were elevated. Additionally, renal values were increased (BUN

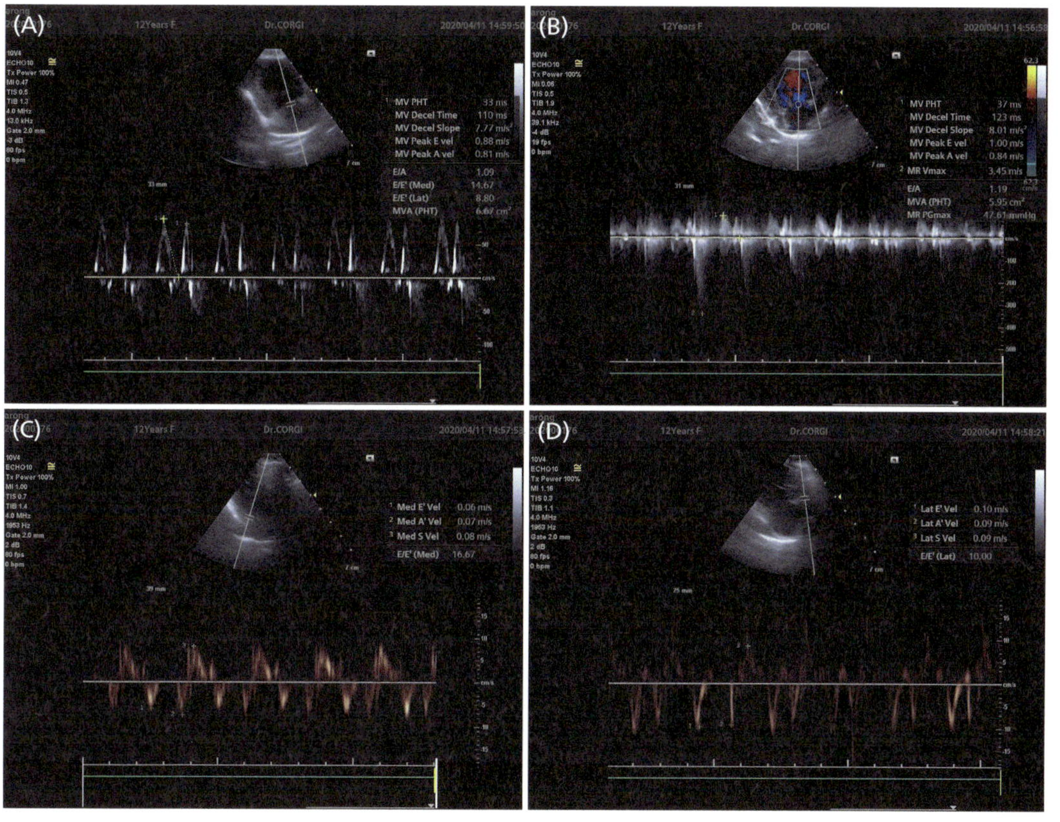

Figure 4-19 ▶ Echocardiography. A: PW Doppler examination showed mitral flow E-peak velocity of 0.88 m/sec. B: CW Doppler examination confirmed mitral regurgitation flow. C & D: TDI Doppler examination showed septal E/E′ 16.67 and lateral E/E′ 10.0, with an average E/E′ of 13, indicating congestive heart failure.

130.8 mg/dL, creatinine 2.48 mg/dL). However, the patient's SDMA level decreased to 28. According to the owner, the patient was breathing more comfortably and had significantly increased vitality and appetite. Also, water intake had slightly decreased, and bowel movement frequency had slightly decreased. Based on the decrease in SDMA level, it was judged that the patient's renal failure had not worsened significantly, but since changes in renal values and electrolyte levels were observed, the torsemide dose was reduced to 0.1 mg/kg, q12h.

The patient returned for a comprehensive cardiac examination 2 weeks later. No significant differences were observed on physical examination, but systolic blood pressure increased significantly to 180 mmHg (measured during a slightly excited state) compared to the previous 120 mmHg. The patient's renal values significantly decreased to BUN 76 mg/dL and creatinine 1.7 mg/dL, but SDMA slightly increased to 29. However, calcium 10.8 mg/dL and phosphorus 4.3 mg/dL returned to normal, and sodium and potassium also remained normal. Thoracic radiographs showed significantly decreased VHS 10.5, VLAS 2.3, RLAS 1.9, and no pulmonary infiltrates were observed (Figure 4-21).

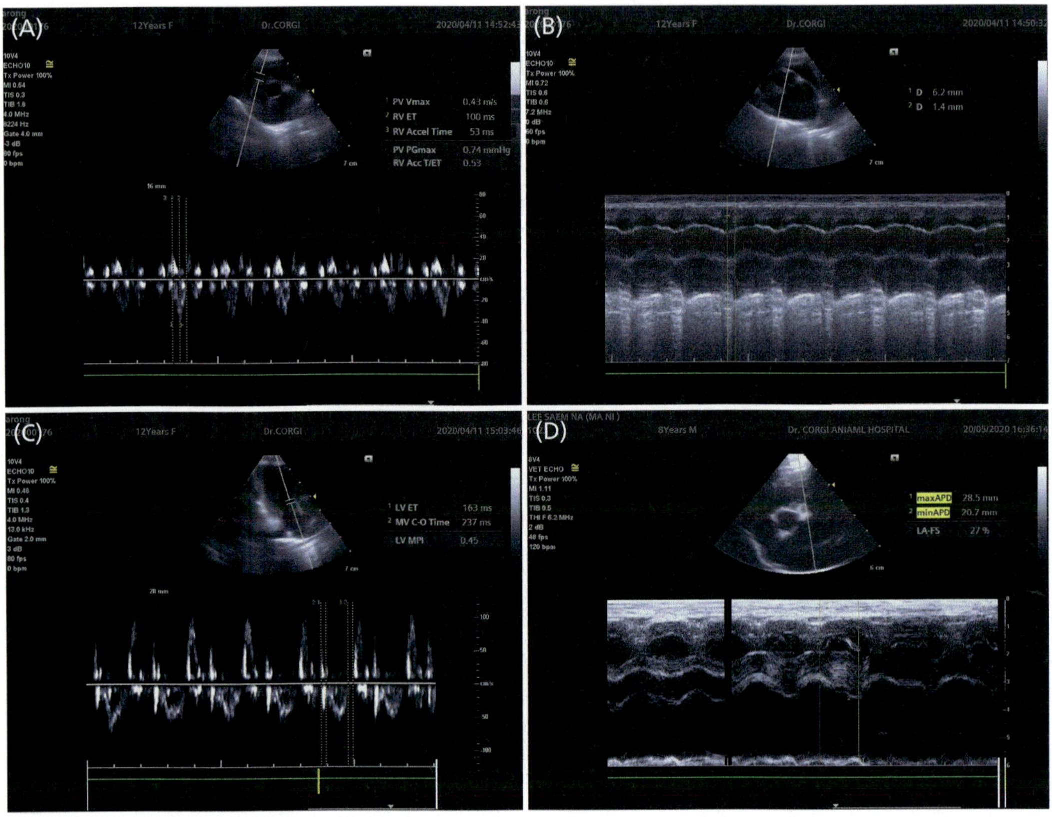

Figure 4-20 ▶ Echocardiography. A: PW Doppler examination of the right ventricular outflow tract flow showed AT 53 m and AT/ET 0.53, with no findings suggesting pulmonary hypertension. B: M-mode aortic short-axis measurement of %RV-FS was 77%, with no findings suggesting pulmonary hypertension. C: PW Doppler examination at the left apical region showed IVRT 35 ms, LVET 163 ms, E/IVRT 2.51, and LV Tei index of 0.45, indicating impairment of both ventricular diastolic and systolic function and suggesting congestive heart failure. D: M-mode aortic short-axis measurement of %LA-FS was 27%, indicating heart failure beyond ACVIM Stage C.

The patient returned for a comprehensive cardiac examination 2 weeks later. No significant differences were observed on physical examination, but systolic blood pressure was stable at 120 mmHg. The patient's renal values decreased to BUN 84 mg/dL and creatinine 1.1 mg/dL, and SDMA decreased to a normal level of 10. Electrolyte tests showed all sodium, potassium, chloride, phosphorus, and calcium levels were normal. Thoracic radiographs showed similar findings to the previous ones, with no significant cardiomegaly or further left atrial enlargement (Figure 4-24).

On echocardiography, LA/Ao Sx 1.67, LVIDd 25.1 mm, LVIDdN 1.93, LVIDd/Ao 2.65, showing a further decrease in left atrial size but a slight increase in left ventricular enlargement (Figure 4-25). However, E/E' (septal 6.99, lateral 8.68), an indirect indicator of left atrial pressure, remained stable at 7.75 (Figure 4-26). %LA-FS increased slightly to 44%, equivalent to ACVIM Stage B. Although a slight increase in left atrial pressure and left ventricular enlargement were observed, the patient's blood test results were all normalized, and

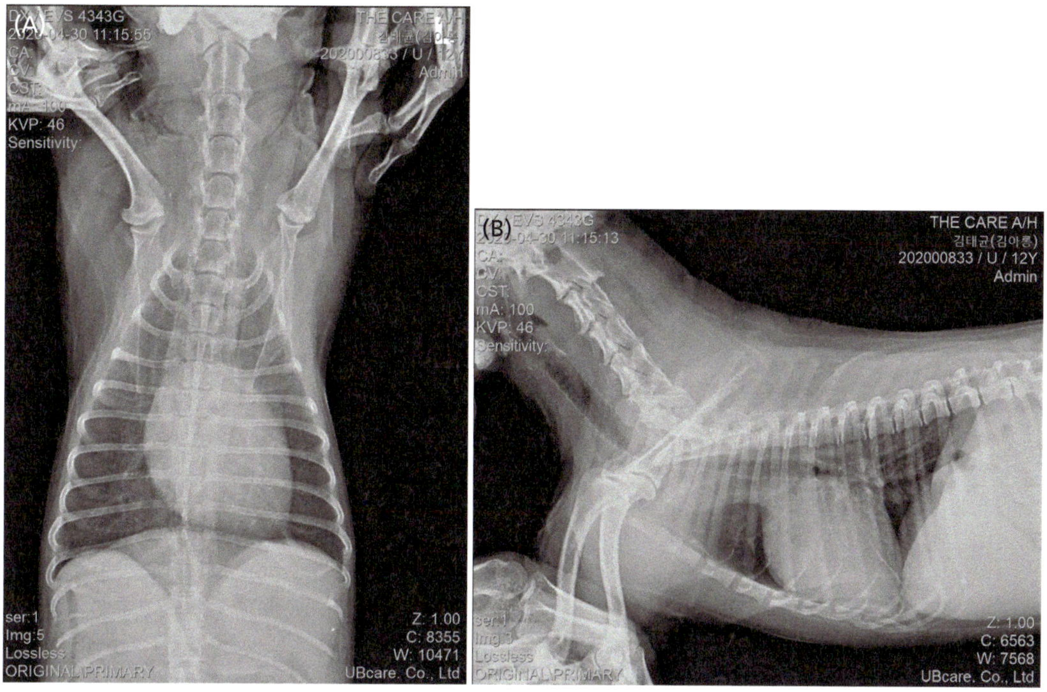

Figure 4-21 ▶ Thoracic radiographs 4 weeks later (A: Ventrodorsal, B: Right lateral). On echocardiography, LA/Ao Sx 1.74, LVIDd 20.3 mm, LVIDdN 1.56, LVIDd/Ao 2.16, showing a noticeable decrease in left atrial size (Figure 4-22). Additionally, E/E' (septal 5.44, lateral 8.78), an indirect indicator of left atrial pressure, significantly decreased to 6.72 (Figure 4-23). %LA-FS increased to 52%, almost within the normal reference range (indicating resolution of left atrial pressure overload). Based on these findings, the patient's diuretic dosage was reduced by 50%. The patient was prescribed torsemide (0.15 mg/kg, q24h), enalapril (0.5 mg/kg, q12h), spironolactone (1 mg/kg, q12h), and pimobendan (0.3 mg/kg, q12h) with a 50% reduction in diuretic dosage. As before, electrolyte and renal function tests were instructed for 2 weeks later to monitor for hypokalemia or hyponatremia due to torsemide, daily sleeping respiratory rates were recorded to check for recurrence of pulmonary edema, and thoracic radiography at a local hospital was instructed 2 weeks after medication. Two weeks later, thoracic radiographs showed no pulmonary infiltrates, the patient's sleeping respiratory rate was more stable at 17-21 breaths/minute, and electrolyte and renal values had almost stably decreased, with SDMA level decreasing to 17. According to the owner, the patient showed almost previous levels of vitality and appetite, and had no problems with defecation or urination. Judging that the patient's condition had not significantly worsened, the patient's medication was maintained for another 2 weeks.

no specific clinical symptoms were observed, so the patient was maintained on the same prescription.

Case Discussion

This case demonstrates how to manage a patient with acute chordal rupture-induced pulmonary edema after rescue. Although the patient's pulmonary edema symptoms were alleviated with high-dose diuretic treatment (furosemide 3 mg/kg, q8h), this case developed acute cardio-renal syndrome due to excessive diuretic treatment. It also represents a case

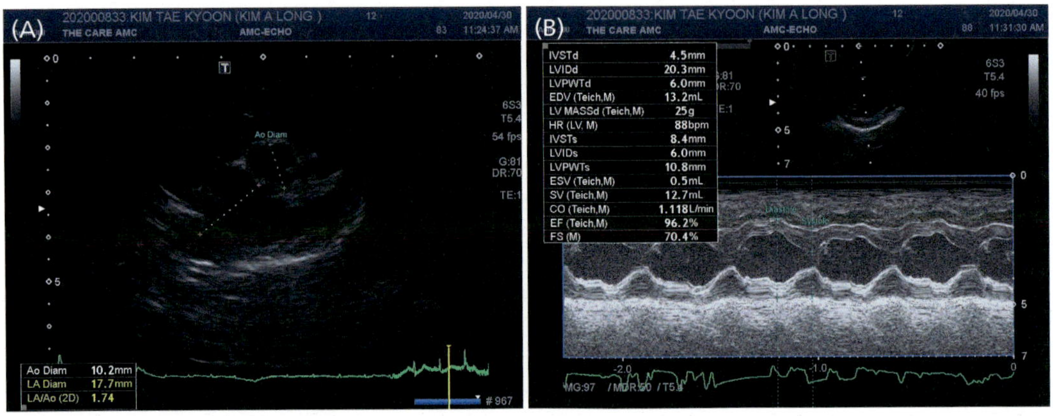

Figure 4-22 ▶ Echocardiography 4 weeks later. A: LA/Ao was measured at 1.74 on the aortic short-axis view. B: M-mode left ventricular papillary muscle short-axis view showed LVIDd of 20.3 mm, LVIDdN of 1.56, and LVIDd/Ao of 2.16.

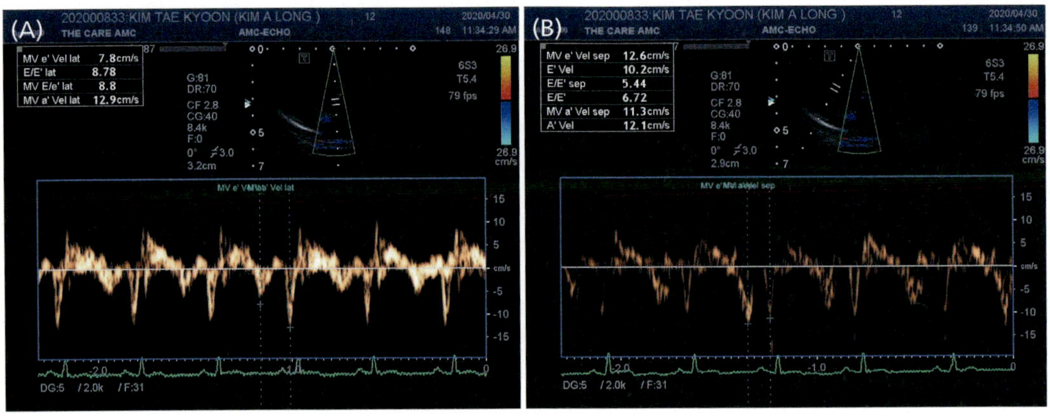

Figure 4-23 ▶ Echocardiography. A&B: TDI Doppler examination showed septal E/E' 5.44 and lateral E/E' 8.78, with an average E/E' of 6.72, indicating that left atrial pressure has fallen to normal levels.

where furosemide tolerance due to a ceiling effect occurred after long-term high-dose furosemide treatment. Furthermore, it illustrates how useful indicators of left atrial pressure (LAP) such as E/E' and %LA-FS can be for patient monitoring.

The patient presented with IRIS Stage II chronic kidney disease and showed loss of appetite and vitality due to excessive diuretics. Nevertheless, echocardiographic indicators showed ACVIM C (ISACHC IIIa) stage heart failure, with prominent excessive left atrial enlargement and increased left atrial pressure. On LV Tei index, the patient showed problems in both systolic and diastolic function. Therefore, drug prescription was given according to ACVIM C stage patients. However, considering the patient's renal status, the diuretic was replaced with torsemide, which is more potent and shows almost no ceiling effect, and the total dose was significantly reduced to 0.3 mg/kg/day (equivalent to furosemide 3 mg/

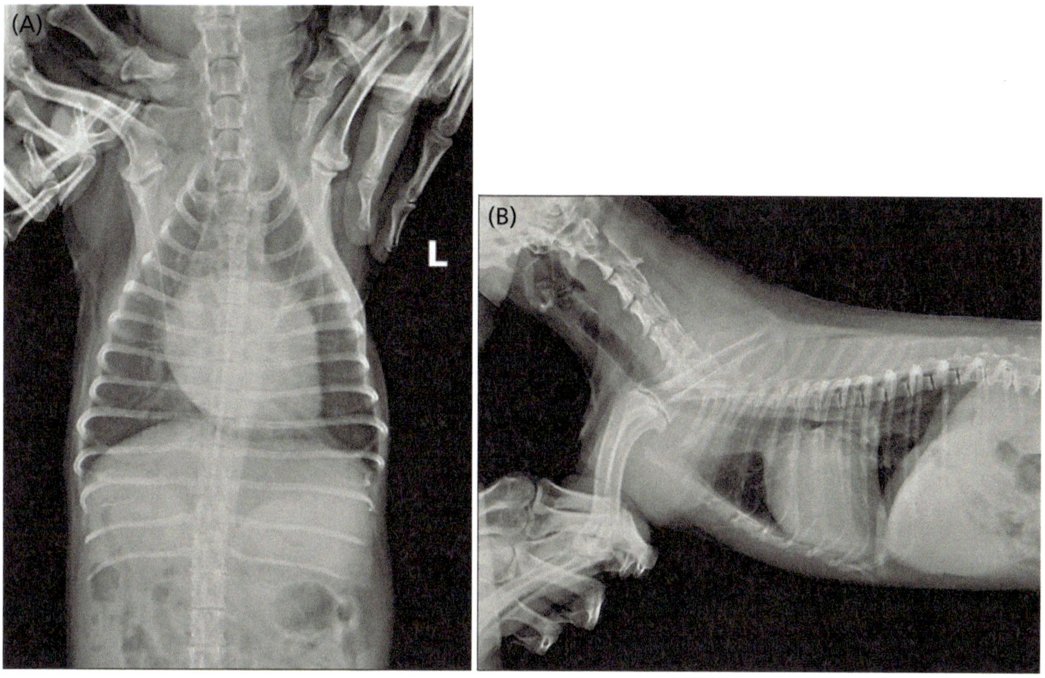

Figure 4-24 ▶ Thoracic radiographs 4 weeks later (A: Ventrodorsal, B: Right lateral). Blue (VHS long axis), orange (VHS short axis), red (VLAS), black (RLAS).

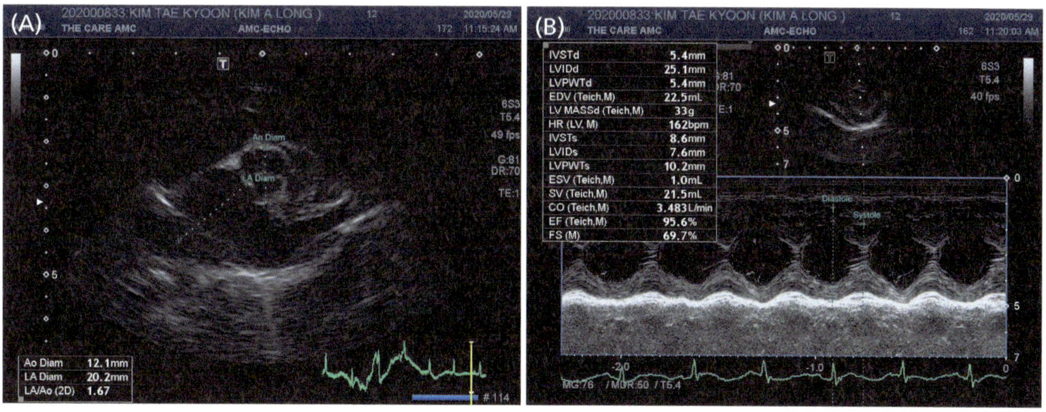

Figure 4-25 ▶ Echocardiography 8 weeks later. A: LA/Ao was measured at 1.74 on the aortic short-axis view. B: M-mode left ventricular papillary muscle short-axis view showed LVIDd of 20.3 mm, LVIDdN of 1.56, and LVIDd/Ao of 2.16.

kg/day). Spironolactone was additionally administered to counteract aldosterone escape and increased left atrial volume load due to diuretic reduction. Torsemide, due to its potent and consistent efficacy, can sometimes cause hypokalemia, hyponatremia, and azotemia due to excessive dehydration within 2 weeks of administration. This phenomenon is more pronounced in patients who have received furosemide for a long time. Furosemide can show a

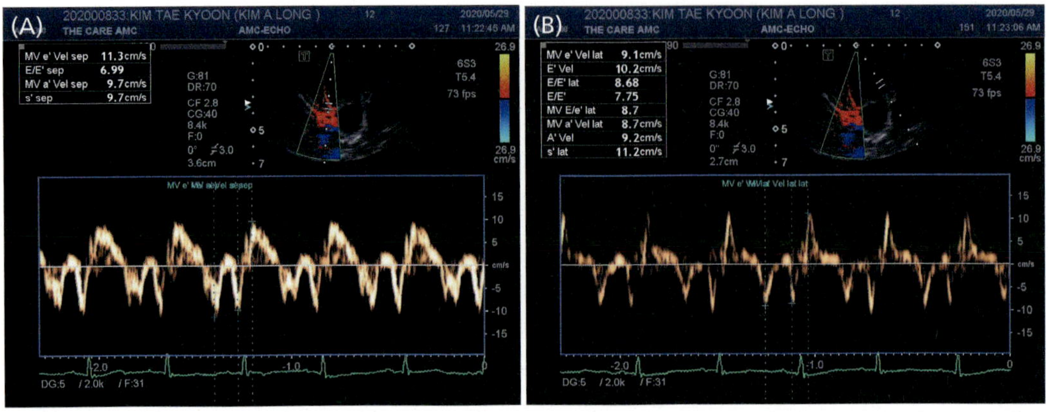

Figure 4-26 ▶ Echocardiography 8 weeks later. A&B: TDI Doppler examination showed septal E/E′ 5.44 and lateral E/E′ 8.78, with an average E/E′ of 6.72, indicating that left atrial pressure has fallen to normal levels.

ceiling effect after 2 weeks of administration, leading to a tendency to administer more than the actually needed diuretic amount. In this patient's case, these side effects were particularly prominent because she had been treated with high-dose diuretics for acute pulmonary edema and patient management. Therefore, despite the patient's LA/Ao Sx of 2.5, left ventricular enlargement was not prominent, so the torsemide dose was significantly reduced to 0.3 mg (equivalent to furosemide 3 mg). However, by having electrolyte levels, renal values, sleeping respiratory rate, and thoracic radiographs checked at a nearby local hospital, the patient's diuretic dosage could be further adjusted. The patient's diuretic dose (torsemide) could be continuously reduced based on these values, in the order of 0.2 mg/kg/day and then 0.15 mg/kg/day.

As seen in the patient's test data, BUN levels did not accurately reflect the patient's renal status, while creatinine and SDMA levels more accurately reflected the patient's renal failure. In particular, SDMA levels well reflected the patient's vitality and appetite. Usually, patients with acute kidney injury experience sudden decreases in vitality and appetite, and these symptoms rapidly improve when renal failure is resolved.

E/E' is an echocardiographic index reflecting left atrial pressure, and in this case, E/E' at presentation was 13, indicating congestive heart failure, and it decreased to 6.72 (normal <9) with appropriate medication. Also, %LA-FS normalized from 27% (corresponding to ACVIM C & D) to 52%. The patient's LA/Ao was >2.5, indicating severe left atrial enlargement and increased left atrial pressure, but it could be reduced to 1.7 (ACVIM B2) with appropriate medication. However, left ventricular enlargement indicators were almost normal or ACVIM B1 stage at presentation but expanded to ACVIM B2 stage during follow-up. These findings are commonly seen when acute pulmonary edema occurs due to sudden chordal rupture in patients with mild MMVD. Cardiac remodeling involves left ventricular enlargement following left atrial enlargement, so in such patients, severe left atrial enlargement with mild left ventricular enlargement is often observed. Similar echocardiographic findings can be seen in patients with acute pulmonary edema due to bacterial endocarditis.

In such patients, indicators of cardiac function are more useful than indicators of cardiac enlargement. In this patient, E/IVRT, LV Tei index, IVRT, and LVET were all closely related to the severity of heart failure and well reflected the patient's condition. This patient's follow-up also demonstrated how drug adjustment according to the patient's heart failure status is possible using indicators like E/E' and %LA-FS.

Pulmonary hypertension is a common complication in MMVD patients. Although this patient showed mild TR findings at the tricuspid valve, subsequent ultrasound examinations did not confirm findings supporting pulmonary hypertension. AT and AT/ET measured at the right ventricular outflow tract (pulmonary artery) were values that could rule out pulmonary hypertension (0.53), and %RV-FS was also 72%, which could rule out pulmonary hypertension. These indicators were very simple and measurable, and were very useful in evaluating the occurrence of pulmonary hypertension in the patient during follow-up.

CASE 4

Patient Information

Patient History: The patient was a 14-year-old spayed female Maltese who had been receiving cardiac care at a local veterinary clinic for 5 years. She was referred due to recurrent pulmonary edema and uncontrolled coughing. Prior to admission, the patient was receiving **pimobendan 0.4 mg/kg q12h; sildenafil 2 mg/kg q12h; furosemide 1.75 mg/kg q8h; benazepril 0.25 mg/kg q12h; tramadol 3 mg/kg q12h; dihydrocodeine 0.5 mg/kg q12h; cyproheptadine 0.2 mg/kg q1h; aluminum hydroxide 20 mg/kg q12h, kremezine 1.5C q12h, and theophylline 12.5 mg/kg q12h**.

Physical Examination

A **grade V-VI/VI systolic regurgitant murmur** was auscultated at the left cardiac apex with palpable precordial thrill. Blood pressure was **120 mmHg systolic**, heart rate was **180 bpm**, respiratory rate was **30 bpm**, body temperature was **38.9°C**, and body weight was **2.2 kg**.

Diagnostic Tests

Laboratory Tests: Key abnormalities in the **CBC** included **anemia (RBC 5.09 M/uL, HCT 35.6%, Hb 12.6 g/dL)**, **phosphorus 4.5 mg/dL**, and **azotemia (BUN 110 mg/dL, creatinine 3.9 mg/dL, SDMA 39 ug/dL)**. Sodium (**150 mEq/L**) and potassium (**4.55 mEq/L**) levels were within normal ranges (suggesting resistance to long-term furosemide use).

Thoracic Radiography: Dorsoventral thoracic radiographs showed **generalized cardiomegaly** with a **VHS of 12V (L=6.5V S=5.5V)**, and mild perihilar infiltration was observed around the heart (Figure 4-27A). The right lateral radiograph (B) revealed cardiomegaly,

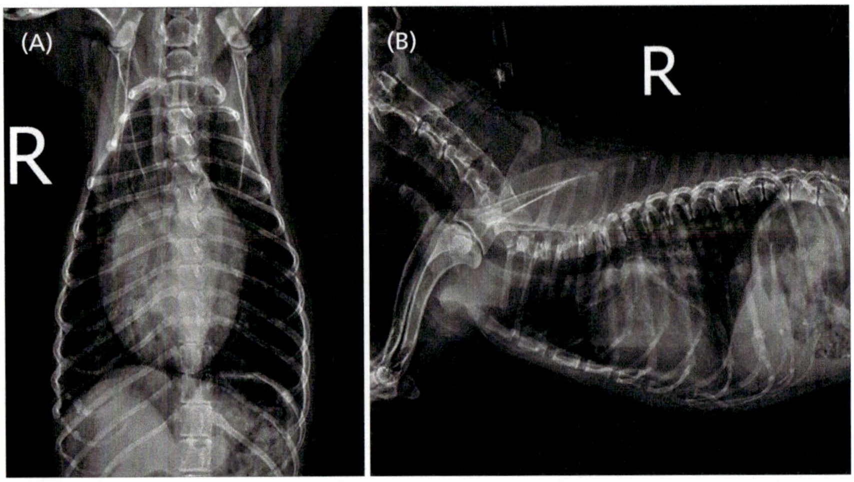

Figure 4-27 ▶ Patient's radiographic findings. Dorsoventral thoracic radiograph showing generalized cardiomegaly with VHS measured at 12V (L=6.5V S=5.5V) and mild infiltration observed around the heart (A). Right lateral radiograph (B) showing cardiomegaly, left atrial enlargement (VLAS 3.2), dorsal tracheal displacement, and a large amount of gas opacity in the stomach due to open-mouth breathing, but no severe hepatomegaly was observed. (A) (B).

left atrial enlargement (VLAS 3.2), dorsal tracheal displacement, and a large amount of gas opacity in the stomach due to open-mouth breathing, but no severe hepatomegaly was observed (Figure 4-27B). The patient's thoracic radiographic findings were consistent with typical observations in **mitral valve insufficiency**.

Echocardiography: On 2-D echocardiography, the **mitral valve appeared abnormally thickened**, and mitral valve prolapse was observed in the apical view, leading to abnormal mitral valve closure and **mitral valve insufficiency**. Additionally, the **left atrium and left ventricle appeared significantly enlarged (LA/Ao 2.41, LVIDd/Ao 2.33, LVIDdN 2.17; Figure 4-28A)**. Color Doppler echocardiography showed mosaic-patterned turbulent blood flow throughout the entire left atrium. CW Doppler echocardiography confirmed **severe regurgitation (5.24 m/s)** from the left ventricle into the left atrium (Figure 4-28B). The E-peak velocity of blood flow from the left atrium to the left ventricle was measured at 1.16 m/s, and the E/e' ratios for the septal and lateral walls were 16.57 and 11.6, respectively. Tricuspid regurgitation was approximately 2.05 m/s, the left ventricle was almost spherical, the pulmonary artery flow AT/ET was 0.52, and no hepatic vein dilation was observed.

Abdominal Ultrasound: No dilation of hepatic or portal veins was observed, and the liver appeared homogeneously echogenic (Figure 4-29A). Mild sludge was observed in the gallbladder. Both kidneys showed **cortical thickening, ill-defined cortico-medullary boundaries, and appeared somewhat atrophied**, with the renal pelvis appearing contracted (Figure 4-29B).

The patient's history of long-term treatment for mitral valve insufficiency and recent admission due to uncontrolled recurrent pulmonary edema and coughing suggested that renal function was compromised, likely due to excessive/prolonged diuretic administration

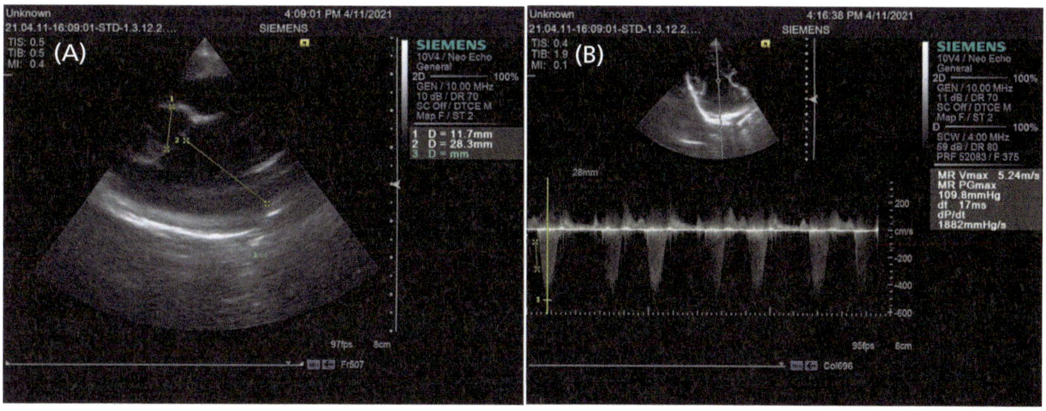

Figure 4-28 ▶ Patient's 2-D echocardiography aortic short axis view (A) and apical view (B). (A) LA/Ao of 2.41 indicates left atrial enlargement consistent with ISACHC Stage III. (B) The regurgitant velocity of mitral valve flow is 5.24 m/s, indicating severe regurgitation, and dP/dt of 1882 mmHg/s, indicating preserved systolic function. (A) (B).

or cardiorenal syndrome. Physical examination revealed a characteristic systolic murmur at the left cardiac apex. Thoracic radiographs showed mild perihilar pulmonary infiltration and cardiomegaly with left atrial enlargement (VHS 12, VLAS 3.2). Echocardiography showed both mitral valve insufficiency and tricuspid valve insufficiency, but only typical left-sided chamber enlargement was observed, with no evidence of right-sided chamber enlargement or pulmonary hypertension. **LA/Ao 2.41, LVIDd/Ao 2.33, LVIDdN 2.17, mitral**

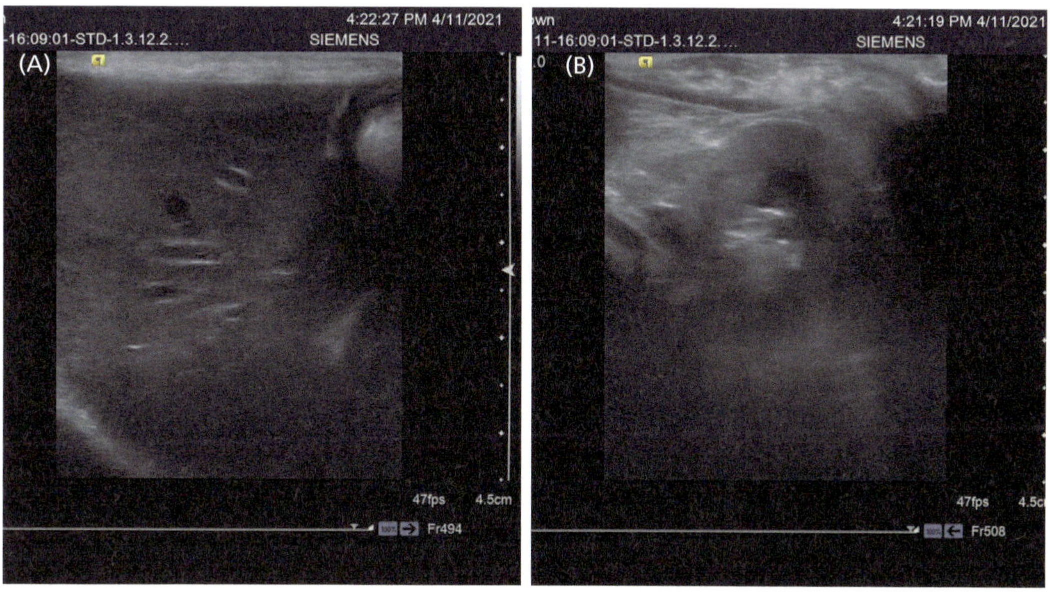

Figure 4-29 ▶ (A) No dilation of hepatic or portal veins is observed, and the liver shows homogeneous echogenicity. (B) Both kidneys show cortical thickening and ill-defined boundaries, appear somewhat atrophied, and the renal pelvis appears contracted. (A) (B).

valve E-peak velocity 1.16 m/s indicated heart failure at ISACHC II-III levels, and septal and lateral E/e' of 16.57 and 11.6 suggested congestive heart failure (values > 12 indicate left atrial pressure > 30 mmHg; normal is < 20 mmHg). Tricuspid regurgitation velocity was approximately 2.05 m/s, the left ventricle was almost spherical, pulmonary artery flow AT/ET was 0.52 (normal < 0.42), and no hepatic vein dilation was observed, which suggests the absence of **pulmonary hypertension**. Abdominal ultrasound showed that both kidneys had cortical thickening, ill-defined boundaries, and a somewhat atrophied appearance, characteristic of chronic kidney disease. Blood tests showed **renal values consistent with IRIS Stage III CKD** (BUN 110 mg/dL, creatinine 3.9 mg/dL, SDMA 39 ug/dL; phosphorus was 4.5 mg/dL, which was normal). Anemia (RBC 5.09 M/uL, HCT 35.6%, Hb 12.6 g/dL) in CBC was a common finding in patients with chronic kidney disease or end-stage heart disease. Notably, despite high-dose diuretic use, sodium and potassium concentrations were **150 mEq/L and 4.55 mEq/L**, respectively, remaining within the normal range. This indicated resistance to long-term furosemide use. Furthermore, in this patient's case, it is possible that phosphorus levels remained normal due to the use of oral phosphate binders. Additionally, the high-dose sildenafil use might have **underestimated** the actual tricuspid regurgitation velocity or right heart enlargement.

Treatment and Patient Observation

In this patient's case, treatment was initiated with the assumption that both heart failure and kidney failure might have been exacerbated by the relatively large number of medications administered to treat both conditions simultaneously. First, **diuretic resistance** was suspected, so **torsemide 0.2 mg/kg q12h** was substituted for furosemide. Additionally, **pimobendan 0.6 mg/kg, enalapril 1 mg/kg, and spironolactone 2 mg/kg** were administered daily. Sildenafil was discontinued because it is known to worsen left heart failure by increasing blood flow returning to the pulmonary veins when administered to mitral valve insufficiency patients without specific right heart enlargement or pulmonary hypertension. Theophylline, tramadol, and dihydrocodeine, administered to control coughing, were replaced with **butorphanol syrup (0.2 mg/kg, PRN)**. Bronchodilators like theophylline can cause palpitations and tachycardia, potentially worsening tachyarrhythmias in heart failure patients.

At a recheck 20 days later, the patient's **heart size had significantly decreased (VHS 12 → 10.5, VLAS 3.2 → 2.6)**, and all mild perihilar infiltration around the heart had resolved. The patient rarely coughed, had good vitality, and the sleeping respiratory rate (SRR) was well maintained at less than 20 breaths/minute. However, the patient's **kidney values (BUN 110 → 130 over mg/dL, creatinine 3.9 → 3.7 mg/dL, SDMA 39 → 59 ug/dL)** and **phosphorus level (4.5 → 6.9 mg/dL)** had slightly worsened. With the diuretic change, the sodium concentration started showing mild hyponatremia again (150 mEq/L to 144 mEq/L). This indicated that diuretic resistance was no longer observed. Although the patient's heart size decreased and clinical symptoms improved, the kidney values increased further, so the diuretic was changed to once-daily administration (morning medication: **torsemide 0.2 mg/kg, pimobendan 0.6 mg/kg, enalapril 1 mg/kg, spironolactone 2 mg/kg; evening medica-**

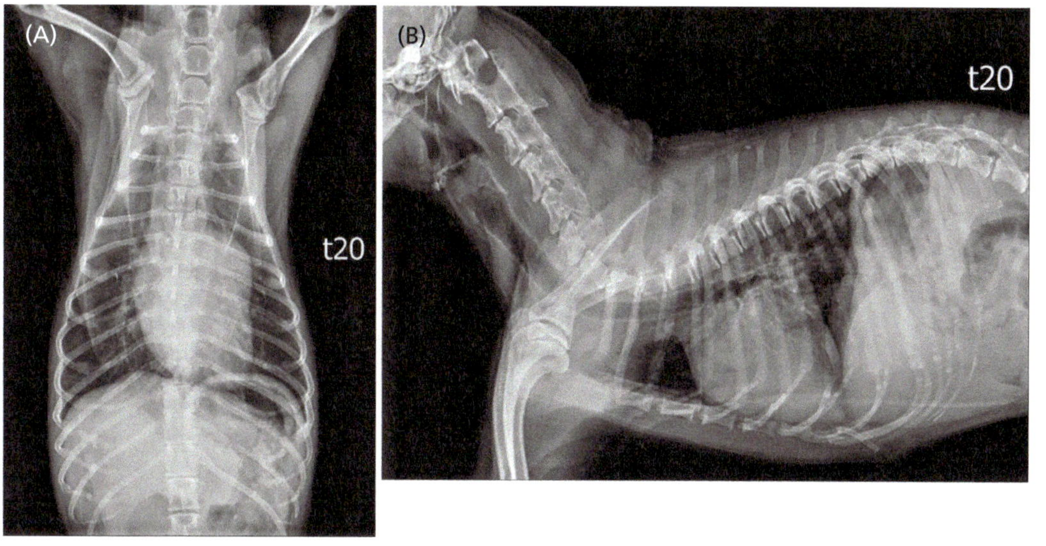

Figure 4-30 ▶ Thoracic radiograph 20 days after treatment. The patient's heart size significantly decreased (VHS 12 → 10.5, VLAS 3.2 → 2.6), and mild perihilar infiltration around the heart is no longer visible. (A) (B).

tion: pimobendan 0.6 mg/kg, enalapril 1 mg/kg).

One week later, the patient re-presented, and a full re-examination was performed. The patient showed no signs of pulmonary edema, but the heart and atria began to enlarge again (VHS 11, VLAS 2.8), and mild coughing (especially when excited) reappeared. Echocardiography findings (LA/Ao 1.85-2.07, %LA-FS 32%, LVIDdn 2.01, mitral valve E-peak 0.81, E/e'(m)' 9, E/e'(l)' 8.1) showed no concerning exacerbation of heart failure symptoms. Even after approximately one month off sildenafil, no right heart enlargement was observed (pulmonary artery flow AT/ET 0.45, %RVFS 76%). Diuretic dosage was maintained without increase. To improve kidney values and manage hyperphosphatemia, **kremezine (1 capsule, q12h)** and **lanthanum 15 mg/kg q12h** were added.

At a recheck 2 weeks later, **improvement in kidney values (BUN 130 over → 113 mg/dL, creatinine 3.7 → 1.6 mg/dL, SDMA 59 → 56 ug/dL)** and **resolution of hyperphosphatemia (6.9 → 5.5 mg/dL)** were observed. Sodium concentration remained slightly hyponatremic at 141 mEq/L. Thoracic radiographic heart size and echocardiographic indices remained at similar levels.

At a recheck one month later, the patient's cardiac clinical symptoms remained good, but he presented with severe anorexia and was somewhat depressed. Kidney values had worsened again (**BUN 113 → 130 mg/dL, creatinine 1.6 → 2.1 mg/dL, SDMA 56 → 80 ug/dL**), and hyperphosphatemia had recurred (5.5 → 13.5 mg/dL). Thoracic radiography showed a decreased heart size, and echocardiographic indices had significantly improved to ISACHC II level. Sodium/chloride concentrations were 160 mEq/L and 126 mEq/L, indicating slight dehydration. This suggested excessive diuretic use and represented a typical **CRS (Cardiorenal Syndrome)** state. Therefore, torsemide was discontinued, and **tolvaptan** was

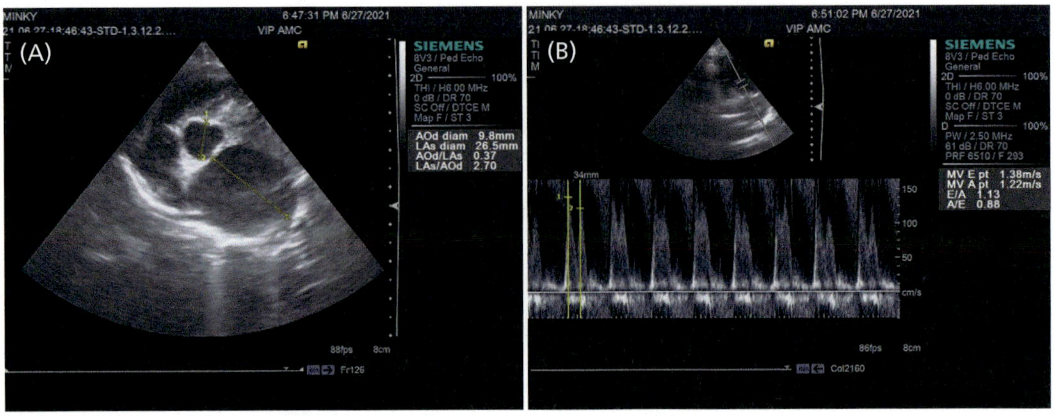

Figure 4-31 ▶ Echocardiography image 3 months after treatment. Echocardiographic indices worsened again to ISACHC III level (LA/Ao 2.7, MV-E peak 1.38 m/s). (A) (B).

initiated (morning: **tolvaptan 7.5 mg/head, pimobendan 0.6 mg/kg, enalapril 1 mg/kg; evening: pimobendan 0.6 mg/kg, enalapril 1 mg/kg**).

At a recheck one week later, the patient's kidney values improved again (**BUN 130 → 119 mg/dL, creatinine 2.1 → 2.7 mg/dL, SDMA 80 → 63 ug/dL**), and hyperphosphatemia resolved (13.5 → 6.5 mg/dL). The patient's dehydration, appetite, and vitality slightly improved (sodium/chloride concentrations were 146 mEq/L and 118 mEq/L).

At a recheck 2 weeks later, kidney values further improved (**BUN 119 → 93.5 mg/dL, creatinine 2.7 → 2.7 mg/dL, SDMA 63 → 34 ug/dL**), and hyperphosphatemia was no longer observed even without phosphate binder administration (6.5 → 4.5 mg/dL). Echocardiographic cardiac indices remained at ISACHC II level: LA/Ao 1.8, LVIDdN 1.8, MV E-peak 0.78 m/s, E/e'(m)' 6.0, E/e'(l)' 6.0. The tolvaptan dose was maintained, while pimobendan was reduced to 0.45 mg/kg and enalapril to 0.75 mg/kg.

At a recheck one month later, the patient's cardiac clinical symptoms remained good, but he presented with severe anorexia and was somewhat depressed. Kidney values had worsened again (**BUN 93.5 → 142 mg/dL, creatinine 2.7 → 3.3 mg/dL, SDMA 34 → 28 ug/dL**), and hyperphosphatemia (4.5 → 7.6 mg/dL) was observed again, along with worsening cardiac indices and intensified coughing. Thoracic radiographs showed no pulmonary infiltration, but cardiomegaly and left atrial enlargement appeared more severe (VHS 12, VLAS 3.4). Echocardiographic indices (Figure 4-31) also worsened to ISACHC III level again (LA/Ao 2.7, LVIDdN 2.2, MV-E peak 1.38 m/s, E/e'(m)' 11.5, E/e'(l)' 12.5). To reduce heart size, tolvaptan administration was increased from once daily to twice daily: **tolvaptan 7.5 mg/head, pimobendan 0.45 mg/kg, enalapril 0.75 mg/kg** for both morning and evening.

The patient had not returned for recheck for two months due to good condition but presented again with severe coughing and increased nocturnal respiratory rate. Kidney values remained at similar levels (**BUN 142 → 89.5 mg/dL, creatinine 3.3 → 3.5 mg/dL, SDMA 28 → 49 ug/dL**), and hyperphosphatemia (7.6 → 8.8 mg/dL) also remained mild, but no severe electrolyte imbalance was observed (Na 147 mEq/L, K 5.69 mEq/L, Cl 116 mEq/L).

However, cardiomegaly and left atrial enlargement on thoracic radiographs appeared more severe (VHS 12.5, VLAS 3.6). Echocardiographic indices also worsened (LA/Ao 2.7 → 2.9, LVIDdN 2.2 → 2.3, MV-E peak 1.38 → 1.29 m/s). Tolvaptan dosage was doubled again, and cardiac medication dosages were also increased (**tolvaptan 15 mg/head, pimobendan 0.6 mg/kg, enalapril 1 mg/kg, q12h**). To manage hyperphosphatemia, **Almagel suspension 100 mg/dose (1.5 mL)** was administered twice daily with meals.

At a recheck one month later, echocardiographic indices improved (LA/Ao 2.9 → 2.53, LVIDdN 2.2 → 2.1, MV-E peak 1.29 → 1.08 m/s, E/e'(m) 12, E/e'(l) 7.71), and kidney values significantly improved (**BUN 89.5 → 41.4 mg/dL, creatinine 3.5 → 2.5 mg/dL, SDMA 49 → 31 ug/dL, phosphorus 8.8 → 4.7 mg/dL**). No electrolyte imbalance other than mild hyperkalemia was observed (K 5.69 → 6.81 mEq/L). The patient had a recheck one month later without further medication increase.

At the recheck one month later, no worsening of clinical symptoms or changes in heart size were observed, but kidney values worsened again (**BUN 41.4 → 119.8 mg/dL, creatinine 2.5 → 4.1 mg/dL, SDMA 31 → 50 ug/dL, phosphorus 4.7 → 10.2 mg/dL**). Electrolyte tests showed that potassium levels normalized (6.81 → 4.98 mEq/L), but mild hyponatremia (143 mEq/L) began to appear. Cardiac medication dosages were not increased. The patient's owner was instructed to administer **subcutaneous fluids (2.5% dextrose/0.45% NaCl solution 75 mL) every other day** at a local clinic and to maximize water intake. Subsequently, kidney values measured at the local clinic every 1-2 weeks showed BUN 88, 257, 207, 120 mg/dL; creatinine 2.6, 3.5, 3.8, 2.9 mg/dL; SDMA 50, 43, 52, 68, 51 ug/dL; and phosphorus 10.2, 4, 11.6, 8.1, 7.1 mg/dL. The patient's kidney values remained at a similar level, neither significantly worsening nor improving.

At a recheck two months later, imaging studies showed improved heart size (LA/Ao 2.34 → 1.95, LVIDdN 2.1 → 1.9, MV-E peak 1.25 → 1.03, E/e'(m) 12 → 6.8), and kidney values remained at similar levels, with no hyperphosphatemia (BUN 154 mg/dL, creatinine 2.8 mg/dL, SDMA 53 ug/dL, phosphorus 4.8 mg/dL). Due to the owner's financial burden and the kidney values not significantly worsening, tolvaptan administration was reduced to once daily, and torsemide was restarted (morning: **torsemide 0.1 mg/kg, pimobendan 0.6 mg/kg, enalapril 1 mg/kg, spironolactone 0.5 mg/kg; evening: tolvaptan 15 mg/head, pimobendan 0.6 mg/kg, enalapril 1 mg/kg**).

At a recheck one month later, all cardiac and renal parameters had worsened, and the patient's anorexia and coughing symptoms intensified (LA/Ao 1.95 → 2.4, LVIDdN 1.9 → 2.1, MV-E peak 1.03 → 0.8 m/s, E/e'(m) 6.8 → 6.15; BUN 154 → 148 mg/dL, creatinine 2.8 → 4.1 mg/dL, SDMA 53 → 69 ug/dL, phosphorus 4.8 → 10 mg/dL). To mitigate the worsening of renal parameters by ACEi, enalapril was removed from the cardiac medication. The medication regimen was simplified: morning: **torsemide 0.1 mg/kg, pimobendan 0.3 mg/kg, tolvaptan 15 mg/head**; evening: **pimobendan 0.5 mg/kg, tolvaptan 15 mg/head**, increasing the tolvaptan dosage.

At a recheck one month later, the patient had an additional history of emergency presentation to a local clinic with apnea and syncope. Pulmonary artery regurgitation (2.5 m/s), tricuspid regurgitation (3.19 m/s), and right heart enlargement were additionally ob-

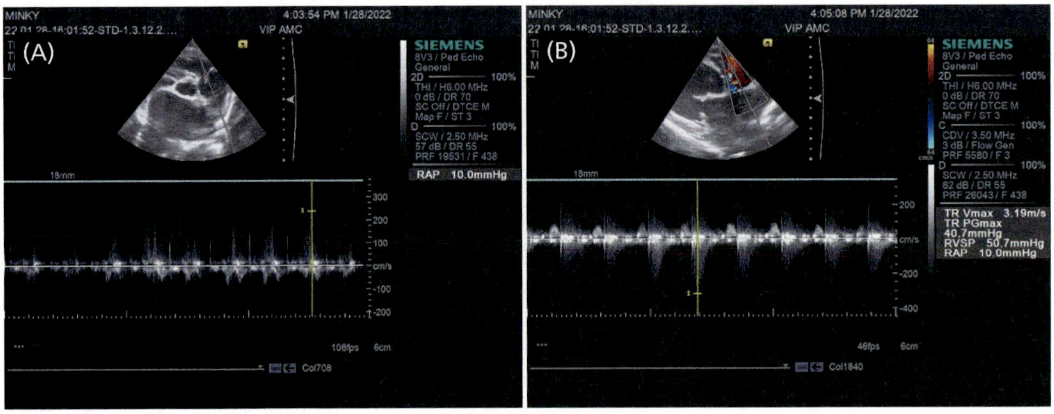

Figure 4-32 ▶ Echocardiography revealed moderate pulmonary hypertension and right heart enlargement along with pulmonary artery regurgitation jet(2.5 m/s) and tricuspid regurgitation jet(3.19 m/s).

served (Figure 4-32). However, kidney values appeared slightly improved (BUN 148 mg/dL, creatinine 4.1 mg/dL, SDMA 69 ug/dL, phosphorus 10 mg/dL). The patient began to show hepatomegaly and diarrhea due to pulmonary hypertension. **Sildenafil 1 mg/kg q12h** was added to the existing cardiac medications for pulmonary hypertension treatment, and **loperamide 0.08 mg/kg** was prescribed for 2 days to relieve diarrhea. The owner was recommended to add psyllium husk (Metamucil, dietary fiber) to the diet to help alleviate loose stools.

One week later, a phone check revealed that the patient was in good condition and diarrhea symptoms had resolved. Due to financial concerns, the patient did not return for further visits and received the same prescription from the local clinic. Two months later, the patient presented to the local clinic as an emergency with apnea accompanied by syncope but could not be resuscitated and passed away.

Case Discussion

Case 13 in this book has already described a case where a cardiorenal syndrome patient was treated using peritoneal dialysis. As already mentioned, the most effective treatment for cardiorenal syndrome is prevention, thus establishing an appropriate heart failure treatment strategy tailored to the patient's heart failure progression state and, most importantly, **optimizing diuretic dosage** is crucial. Such strategies include:

Administering smaller, fractionated doses rather than large, single doses (BID → TID).

Combining with other diuretics or switching to a different diuretic if resistance to one diuretic is observed.

Adding other heart failure medications (e.g., spironolactone, nitrate patch, amlodipine) to reduce diuretic usage.

Since **Cardiorenal Syndrome (CRS)** in heart failure patients mostly arises from inappropriate diuretic treatment and reduced renal perfusion due to hypotension, methods such as administering **maintenance subcutaneous fluids** to prevent excessive dehydration and **increasing inotropic agent dosages** can also be employed. In such cases, 0.45% NaCl with 2.5% dextrose solution can be given subcutaneously at maintenance doses (30 mL × body weight + 70 mL) divided into two administrations. However, in patients with severe congestive heart failure, even this dose can induce pulmonary edema, so respiratory rate monitoring is necessary. In actual clinical settings, peritoneal dialysis can be too labor-intensive, and excessive subcutaneous fluids can cause pulmonary edema, necessitating an alternative to loop diuretics like furosemide. **Tolvaptan**, as a V2-receptor antagonist, can exert diuretic effects with minimal impact on the kidneys.

According to a study published by Onogawa et al. in 2011, administering **tolvaptan 10 mg/kg** to dogs with congestive heart failure induced diuresis without significant electrolyte imbalances or adverse renal effects, and cardiac hemodynamic parameters significantly improved. However, the furosemide dose used for comparison in that study (10 mg/kg) is not used in actual clinical practice, so the appropriate dosage in dogs has not yet been established. Although the study reported a lower diuresis compared to the 10 mg/kg furosemide group, it did not clearly research what dose could provide the minimum diuretic effect or how to adjust the dosage. However, a study by Miyazaki et al. in 2007 comparing the efficacy of tolvaptan and furosemide in dogs showed a **furosemide to tolvaptan efficacy ratio of 1:3.3 (Figure 4-33)**, and diuretic effect increased with increasing dosage.

Based on these research findings, the author began administering tolvaptan in clinical practice. When it was time to increase furosemide, tolvaptan dosage was adjusted according to the expected furosemide dose. However, **tolvaptan tablets cannot be crushed as this**

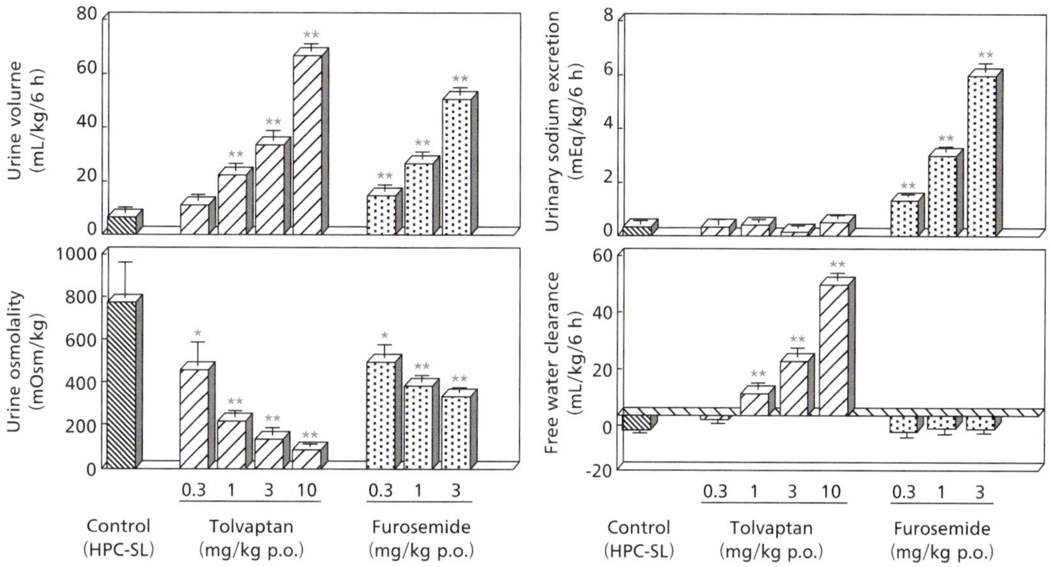

Figure 4-33 ▸ Diuretic effect of tolvaptan in dogs (Miyazaki et al, 2007) (Graph showing urine volume, urinary sodium excretion, urine osmolality, and free water clearance responses to varying doses of Tolvaptan and Furosemide in dogs).

completely negates their effect, so 15 mg or 30 mg tablets were divided into two to match the estimated dose and administered once or twice daily. In the actual patient in this case, torsemide 0.2 mg/kg (equivalent to furosemide 2 mg/kg) was needed to control heart failure symptoms, which would correspond to a tolvaptan dose of 15 mg (2.2kg body weight × 2 mg/kg furosemide × 3.3 = 14.52). However, considering the patient's cardiac status, hydration status, and renal status, an initial dose reduced by half, **7.5 mg (1/2 of a 15 mg tablet)**, was tried. The results were satisfactory, so that dose was maintained, and tolvaptan dosage was increased according to the method of increasing furosemide whenever diuretic increase was necessary. This strategy was effective in controlling CRS in the patient. However, in other cases, long-term use of tolvaptan has shown tolerance, and in such instances, a temporary discontinuation of tolvaptan has helped restore its diuretic effect. To date, tolvaptan has been administered regardless of food intake, and there has been experience of concomitant administration with antibiotics or antacids, with no particular problems identified.

In heart failure patients, it is common to see an increase in serum creatinine levels with long-term use of ACE inhibitors (ACEi) like enalapril. This phenomenon occurs because ACEi reduce the pressure in the efferent renal arterioles, leading to a decrease in intraglomerular pressure, a common side effect in renal failure patients. Although various prospective and retrospective studies on ACEi in heart failure patients show that the positive effects of ACEi on cardiac and renal function outweigh the negative ones, the author sometimes temporarily reduces or discontinues ACEi or uses ARBs such as telmisartan (Semintra) or valsartan instead. However, in clinical practice, unexplained increases in renal values have occasionally been observed when Semintra was administered to asymptomatic heart failure patients, so this strategy is no longer used. Instead, ACEi or ARBs are simply reduced or discontinued until renal values improve. This strategy was also used in this case and somewhat helped improve CRS symptoms.

Tolvaptan

Tolvaptan, a V2−receptor antagonist, is being utilized as a novel therapeutic agent for **hyponatremia** in heart failure patients. Specifically, tolvaptan, marketed under the brand name Samsca, is used as a selective, competitive vasopressin receptor 2 antagonist to treat hyponatremia associated with congestive heart failure, liver cirrhosis, and the syndrome of inappropriate antidiuretic hormone (SIADH). Research suggests that in dogs with congestive heart failure, tolvaptan can be used as a new concept diuretic that reduces cardiac preload without adversely affecting renal function, systemic hemodynamics, or the neurohormonal system. In animal models of human **polycystic kidney disease (PKD)**, tolvaptan has been shown to significantly reduce the volume of renal cysts and fibrosis. Furthermore, recent clinical trials (ACTIV in CHF) for the treatment of congestive heart failure in humans have revealed that adding tolvaptan to existing medications improves edema and hyponatremia without impacting blood pressure, heart rate, or renal function. This case describes the successful management of **Cardiorenal Syndrome (CRS)** in a dog that developed the condition due to excessive diuretic use, treated with tolvaptan.

CASE 5

Patient Information

Patient: 12-year-old neutered male Shih Tzu (weight 5.4 kg).
Patient History: The patient was diagnosed with **MMVD ISACHC IIIA (ACVIM C)** and had been managed for approximately 11 months. He presented as an emergency with sudden dyspnea and coughing as chief complaints. At the time, the patient was taking **torsemide 0.13 mg/kg, pimobendan 0.38 mg/kg, enalapril 0.65 mg/kg, and spironolactone 1.3 mg/kg,** all q12h PO.

Physical Examination and Phonocardiography

On physical examination, a **grade IV/V pansystolic regurgitant murmur** was auscultated at the left cardiac apex. Systolic blood pressure was 85 mmHg, heart rate was 144 bpm, and respiratory rate was 48 breaths/minute. On lung auscultation, **high-pitched fine crackle sounds** were heard in the right lung lobe.

Diagnostic Tests

Laboratory Tests: Blood collection could not be performed due to the patient's poor respiratory status.

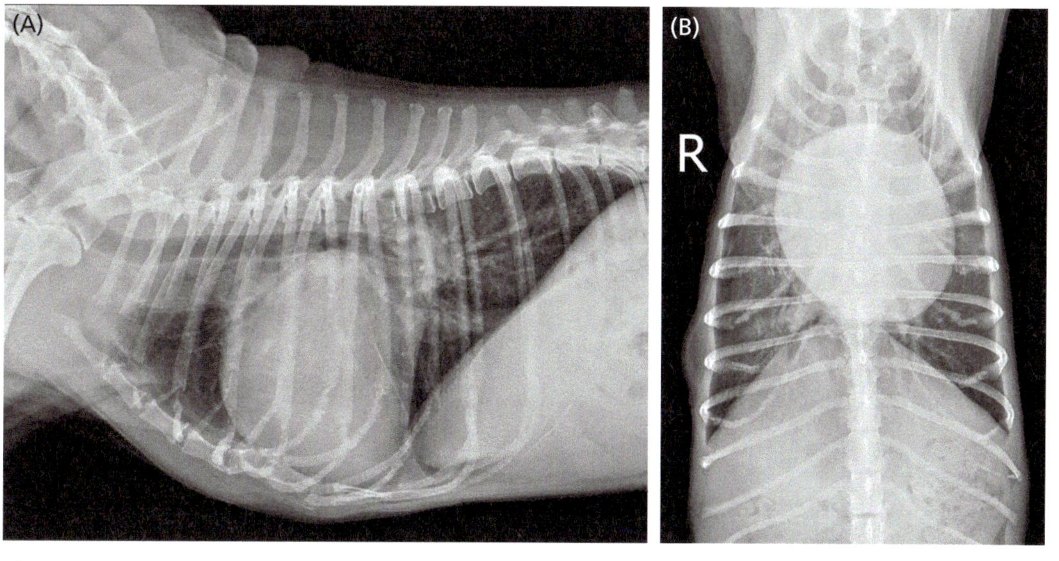

Figure 4-34 ▶ Patient's radiographic findings on admission. A: Lateral thoracic radiograph showing generalized cardiomegaly, marked left atrial enlargement, and main bronchus compression. B: Dorsoventral thoracic radiograph showing a mild globoid-shaped cardiac silhouette and interstitial infiltration in the right caudal lung lobe.

Thoracic Radiography: Immediately after admission, a lateral thoracic radiograph confirmed **generalized cardiomegaly** and marked **left atrial enlargement**, along with main bronchus compression. The dorsoventral thoracic radiograph showed a mild **globoid-shaped cardiac silhouette** and interstitial infiltration in the right caudal lung lobe (Figure 4-34A, B). The patient's thoracic radiographic findings were typical for **cardiogenic pulmonary edema** in a patient with mitral valve insufficiency.

Echocardiography: On 2-D echocardiography (with color Doppler), mosaic-patterned turbulent blood flow was observed throughout the entire left atrium, and **mitral valve prolapse** was confirmed, but a frail movement was not observed, so it was not assessed as a complete chordal rupture. However, compared to the cardiac size and functional indices evaluated one month prior, all were found to have worsened. The **LA:Ao ratio** was measured at 2.79 in the right parasternal short axis aortic view, and the left ventricular end-diastolic dimension was 39.3 mm in the short axis M-mode image at the papillary muscle level, with the **LVIDd:Ao ratio** increased to 3.04. Furthermore, the patient's **E-peak velocity**, measured by PW Doppler, was 1.42 m/s, suggesting excessive left atrial pressure elevation (Figure 4-35).

Emergency Treatment and Additional Tests: Based on imaging diagnostic findings, the patient was diagnosed with **cardiogenic pulmonary edema** in a MMVD ISACHC IIIA stage patient, and admission for treatment was decided. The patient initially received **oxygen supplementation (oxygen cage 5 L/min), furosemide (2 mg/kg IV, then 0.66 mg/kg/hr CRI)**, and **dobutamine (5 ug/kg/min CRI)**. Due to high excitement in the hospital, **butorphanol (0.2 mg/kg IV)** was administered for sedation. Excitement persisted even 30 minutes after admission, and at 8:35 PM, the patient collapsed and had defecation and urination while unconscious. The syncope lasted for approximately 30 seconds, after which breathing returned, but consciousness remained depressed, and the gum mucous membranes were severely pale. ECG monitoring at that time showed a heart rate of 110-120

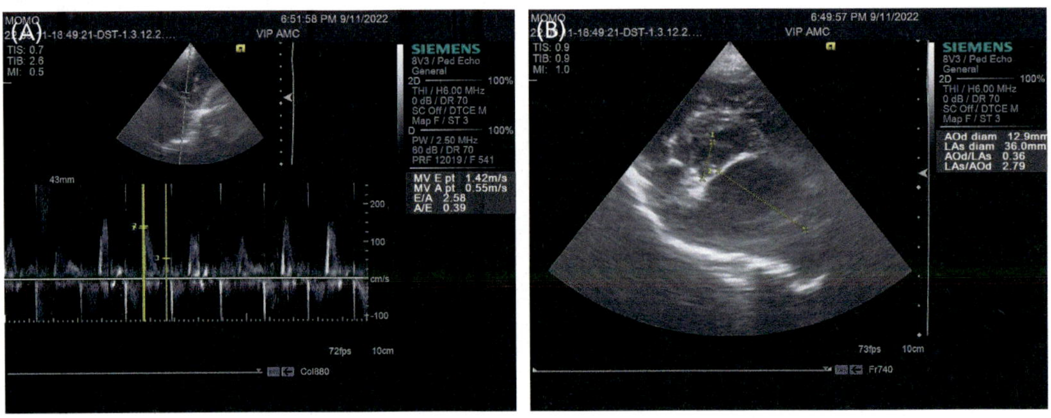

Figure 4-35 ▸ Patient's echocardiographic findings on admission. A: The patient's LA/Ao ratio was 2.79 in the 2-D echocardiography aortic short axis view, showing severe left atrial enlargement. B: E-peak velocity measured by PW Doppler in the left apical long axis view was 1.42 m/s, suggesting excessive left atrial pressure elevation.

bpm, with no significant abnormalities in the ECG waveform. Femoral pulse was not palpable, and systolic blood pressure measured by Doppler sphygmomanometer was 50 mmHg. Horizontal nystagmus was observed for approximately 4 minutes immediately after syncope. Accordingly, the dobutamine CRI rate was doubled, and **dexamethasone (0.1 mg/kg IV)** was additionally administered. Blood pressure, respiratory rate, and body temperature were monitored hourly, and the rates of dobutamine CRI, furosemide CRI, and 5% dextrose fluid were adjusted to a total of 2-2.5 ml/kg/hr. There were no further syncopal events overnight, and heart rate and respiratory rate stabilized, but despite vasopressor treatment, systolic blood pressure remained consistently hypotensive around 100 mmHg.

Upon re-examination of thoracic radiographs the next day, the infiltration in the right lung field was observed to be similar, but overall cardiomegaly was exacerbated, and the cardiac silhouette appeared **globoid-shaped** (Figure 4-36A, B). However, hepatomegaly and pleural/abdominal effusion were not observed. On echocardiographic examination, an **anechoic space** between the pericardium and epicardium was observed, but no mass suggestive of a thrombus was seen. Based on these imaging diagnostic findings, **acute hemopericardium due to left atrial rupture** was diagnosed.

Treatment and Patient Observation

Laboratory tests performed the day after admission revealed **mild anemia (HCT 30.5%; RR 37.3-61.7)**, **mild hyponatremia (142 mmol/L; RR 145-151)**, and **hypochloremia (109 mmol/L; RR 110-119)** due to diuretic treatment. Additionally, **prerenal azotemia (BUN 98 mg/dl; RR 7-27, creatinine 4.0 mg/dl; RR 0.5-1.8)** occurred due to decreased GFR from

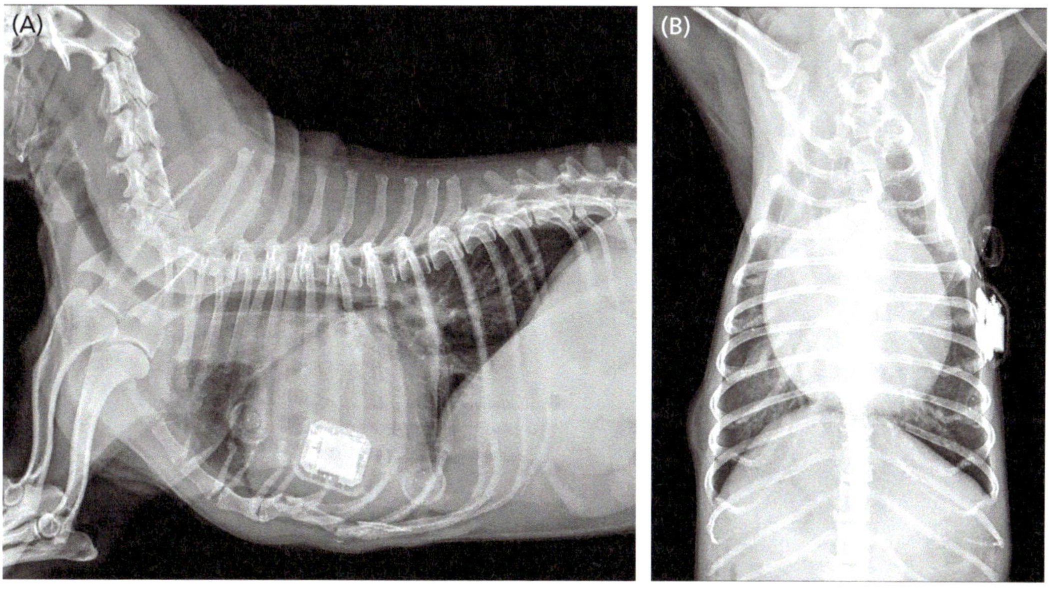

Figure 4-36 ▸ Patient's radiographic findings on the day after admission (A: Lateral view, B: Dorsoventral view). Infiltration in the right lung field is similar, but overall cardiomegaly is exacerbated, and the cardiac silhouette is globoid-shaped.

hypotension, accompanied by **elevated SDMA (SDMA 33 ug/dl; RR 0-18)** and **hyperphosphatemia (Phosphorus 15.2 mg/dl; RR 2.5-6.8)**. Mild elevation in liver enzymes was also confirmed due to presumed ischemic injury (ALP 306 U/L; RR 23-212, ALT 174 U/L; RR 10-100).

Although the patient's hypotension was considered to be influenced by cardiac tamponade due to pericardial effusion, given the patient's unstable respiratory status and the small amount of pericardial fluid, it was judged that reducing left-sided cardiac load through **high-dose diuretic treatment** would be more beneficial for patient recovery than pericardiocentesis. Therefore, conservative therapy focusing on vasopressors and diuretics was implemented. Since the patient was already in a severe state of prerenal azotemia, furosemide CRI was discontinued, and **tolvaptan (30 mg/dog, PO, BID)**, a vasopressin V2 receptor antagonist, was administered instead. In addition, to improve cardiac contractility and maintain blood pressure, **pimobendan (0.3 mg/kg, PO, TID)** was added to the already ongoing dobutamine CRI.

Thoracic radiographs taken 2 days after admission showed a **decrease in heart size (decreased VHS)**, suggesting a reduction in pericardial effusion, so no additional pericardio-

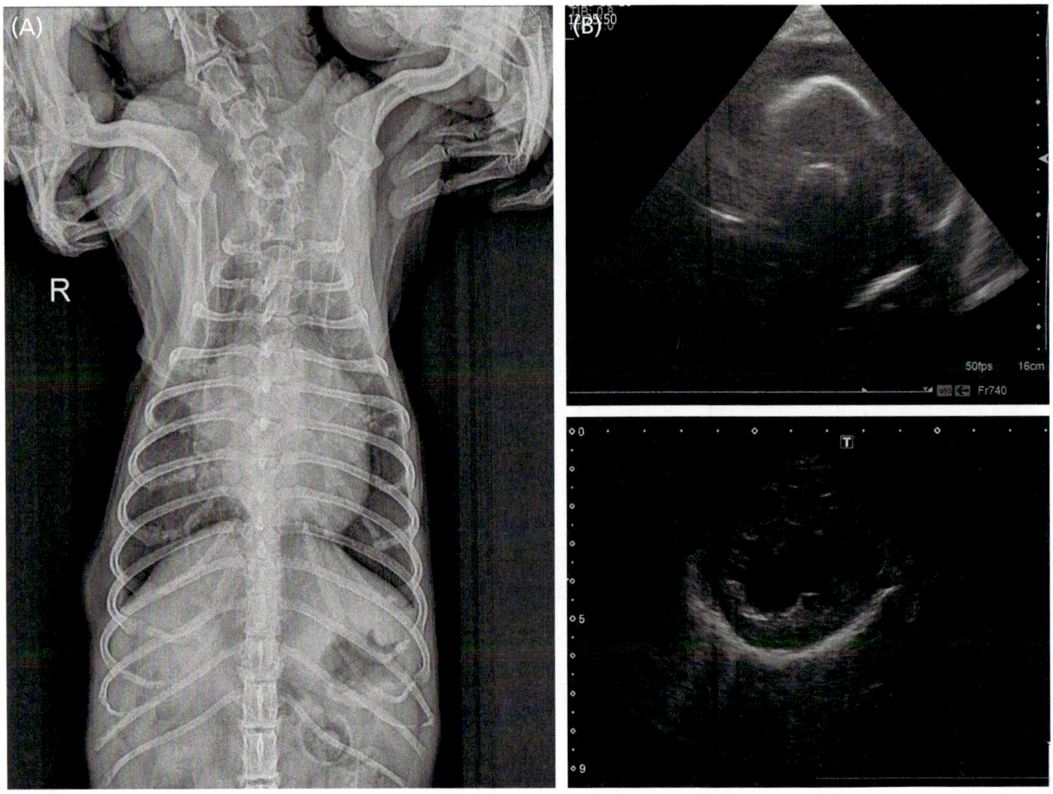

Figure 4-37 ▶ Patient's imaging findings on discharge day (Day 5 of admission). A: Dorsoventral thoracic radiograph. Infiltration in the right lung field has disappeared, and heart size has significantly decreased. B: Echocardiography (2D). Pericardial effusion within the pericardium has significantly decreased on discharge day (B2) compared to the day after admission (B1).

centesis was performed. However, as azotemia worsened and hyperphosphatemia persisted, **sevelamer (200 mg/dog, PO, TID)**, a phosphate binder, was added to the existing prescription. Subsequently, changes in pericardial effusion were monitored every 24 hours through thoracic radiography, and the patient's hydration status was assessed. On day 3 of hospitalization, the patient's vital signs improved, pericardial effusion on thoracic radiographs was confirmed to be decreasing, and overall improvement, including kidney values, was observed. However, a rapid increase in HCT was assessed to be due to blood concentration caused by high-dose diuretic treatment, so the tolvaptan dose was reduced by 50%, and oral electrolyte solution **Ringerlight soln (Kyungnam, Korea) (10 ml/dog, PO, 5 times/day)** was started. On day 5 of hospitalization, systolic blood pressure was well maintained at 120-130 mmHg despite discontinuation of dobutamine CRI. Improvement in lung infiltration, decrease in pericardial effusion on echocardiography, and slight improvement in kidney values were confirmed. Therefore, the patient was discharged with the following medication prescription: **tolvaptan (15 mg/dog, PO, BID), pimobendan (0.3 mg/kg, PO, TID), enalapril (0.5 mg/kg, PO, BID), sevelamer (200 mg/dog, PO, TID)**.

At a recheck 6 days after discharge, the patient's vital signs were good, and systolic blood pressure was measured at 100 mmHg. Blood tests showed no significant abnormalities other than mild anemia and azotemia, and pulmonary infiltration and pericardial effusion were no longer observed. Cardiac size and functional indices on echocardiography at the first recheck after discharge were maintained similarly to those on admission, but rapid progression of cardiomegaly was observed at the second recheck approximately 2 weeks later. Accordingly, the patient was re-evaluated as **MMVD ISACHC IIIB (ACVIM D)**, and subsequent medication therapy proceeded as follows: (morning/evening: **tolvaptan 15 mg/head, torsemide 0.1 mg/kg, pimobendan 0.5 mg/kg, enalapril 1 mg/kg, spironolactone 1 mg/kg; lunch: pimobendan 0.25 mg/kg**), and the patient is still alive and comes for rechecks monthly.

Case Discussion

Left Atrial Rupture: Clinical Considerations and a Case Study

In clinical settings, left atrial rupture is primarily caused by **tumors/inflammation** or **severe left atrial pressure elevation**. In this particular case, it's highly probable that the rapid progression of excessive left atrial pressure elevation due to long-standing mitral regurgitation, coupled with subsequent rapid left atrial enlargement, led to **splitting of the left atrial endocardium**. In such instances, instead of causing rapid hemorrhage, blood slowly leaks from the left atrium into the pericardial sac. This could makes it difficult to observe blood clots within the pericardial fluid. In one study, 4 out of 11 dogs with left atrial rupture showed clot in the pericardial effusion under echocardiographic examination. If pericardiocentesis was performed, a dark, wine-colored bloody effusion, similar to general hemorrhagic pleural effusion, can be observed. However, a decrease in the overall circulating blood volume can lead to a drop in blood pressure, and excessive pericardial fluid causing cardiac tamponade can result in **hypotensive shock and tachycardia**. This patient, upon emergency admission, presented with severe hypotension and tachycardia.

Typically, in cases of acute left atrial rupture with concomitant hemorrhage, **blood transfusions and pericardiocentesis** may be indicated. However, in cases like this one, where the cause is endocardial splitting due to increased left atrial pressure, the pericardial fluid accumulates slowly, and tamponade is not severe. In these situations, increasing dosages of **diuretics and inotropes** to reduce left atrial pressure is relatively safer and more effective than pericardiocentesis. In this patient's case, echocardiography did not show pericardial fluid sufficient to cause cardiac tamponade, so the patient was recovered with inotropes, diuretics, and oxygen therapy rather than pericardiocentesis.

In cases of acute left atrial rupture with rapid blood extravasation, cardiac arrest often occurs during pericardiocentesis due to hypotension and tachycardia. In such scenarios, it's crucial to **maintain blood pressure with transfusions and inotropes** and stabilize the patient before attempting to remove the pericardial fluid.

Pericardial Effusion Characteristics

Generally, pericardial effusions caused by severe right heart failure are often transudates (due to decreased colloid osmotic pressure) or modified transudates. Conversely, idiopathic hemorrhagic pericardial effusions are frequently characterized by a distinct wine-colored, severely bloody exudate without underlying cardiac disease, and the blood in the pericardial fluid typically does not clot. In cases of neoplastic effusion, specific tumor cells may or may not seen on cytology, and the fluid is usually a modified transudate. However, if it's an angiosarcoma, severe hemorrhagic findings may be present. Furthermore, mesotheliomas, which cause malignant effusions, can be hemorrhagic but show numerous mesothelial cells with marked nuclear atypia, including multinucleation, nuclear pleomorphism, high nuclear-to-cytoplasmic ratio, and coarse chromatin.

In this patient, the pericardial fluid exhibited typical hemorrhagic characteristics, and since it was not a case of acute massive hemorrhage, no thrombi or blood clots were observed within the pericardial fluid. All observed leukocytes in the pericardial fluid were intact, resembling those seen in blood. Therefore, hemorrhage due to left atrial rupture was the most suspected diagnosis.

Management of Renal Dysfunction in Left Atrial Rupture

This patient had been undergoing treatment for chronic heart failure for a prolonged period, resulting in some degree of kidney damage. The sudden left atrial rupture, accompanied by hypotension, significantly worsened renal function. Shortly after administering loop diuretics to reduce left atrial pressure, severe azotemia developed. Consequently, to prevent further kidney damage, **tolvaptan** was added to reduce left atrial pressure, as it has a relatively lower impact on renal function. As previously mentioned, tolvaptan is a vasopressin receptor 2 (V2) antagonist that specifically removes water without electrolyte exchange. After emergency treatment for acute hemorrhage, the patient was stabilized by adding tolvaptan.

To summarize this case, it describes a patient with chronic severe mitral regurgitation who experienced left atrial rupture due to excessive left atrial enlargement. However, because the hemorrhage occurred gradually, the bloody pericardial effusion was successfully managed with conservative medical therapy, including blood pressure maintenance and left atrial pressure reduction, without the need for pericardiocentesis.

CASE 6 Mitral Regurgitation - Transcatheter Edge-to-Edge Mitral Valve Repair (TEER) with V-clamp

Patient Information

Patient: 10-year-old intact female Maltese (3.5 kg body weight)

History and Physical Examination

The patient was referred after being diagnosed with ACVIM B2 stage mitral regurgitation and pulmonary hypertension at a local clinic. She had been managed with pimobendan and enalapril for 7 months. Her heart failure suddenly worsened, and torsemide (0.1 mg/kg, q24h) was added to her regimen. Upon presentation, a grade IV/VI holosystolic regurgitant murmur was auscultated at the left apex. Systolic blood pressure was 150 mmHg, heart rate was approximately 200 bpm, respiratory rate was 27 breaths per minute, and aortic pulses showed a pulsus alternans pattern.

Laboratory Findings

No significant abnormalities were noted on laboratory tests.

ECG Findings

The electrocardiogram showed a sinus rhythm with a heart rate of 175-205 beats per minute, indicating sinus tachycardia.

Radiographic Findings

Thoracic radiographs revealed cardiomegaly (VHS 12.2) with clear left atrial (VLAS 3.1) and left ventricular enlargement. While there was no pulmonary infiltrate, significant bronchial compression at the tracheal carina and loss of ventral bend with elevation were evident (Figure 4-38).

Echocardiographic Findings

The LA/Ao SX was measured at 1.99 (ISACHC II), and LVIDd was 30.8 mm, with LVIDd/Ao at 2.61 (ISACHC III) (Figure 4-39 A&B). Pulsed-wave Doppler showed a mitral inflow E-peak velocity of 1.11 m/sec (ISACHC II). Continuous-wave Doppler confirmed high-velocity mitral regurgitation flow. Tissue Doppler imaging revealed septal E/E' of 8.54 and lateral E/E' of 10.09, with an average E/E' greater than 9, suggesting elevated left atrial pressure (Figure 4-39 C&D). Color Doppler revealed mosaic flow regurgitating into the left atrium.

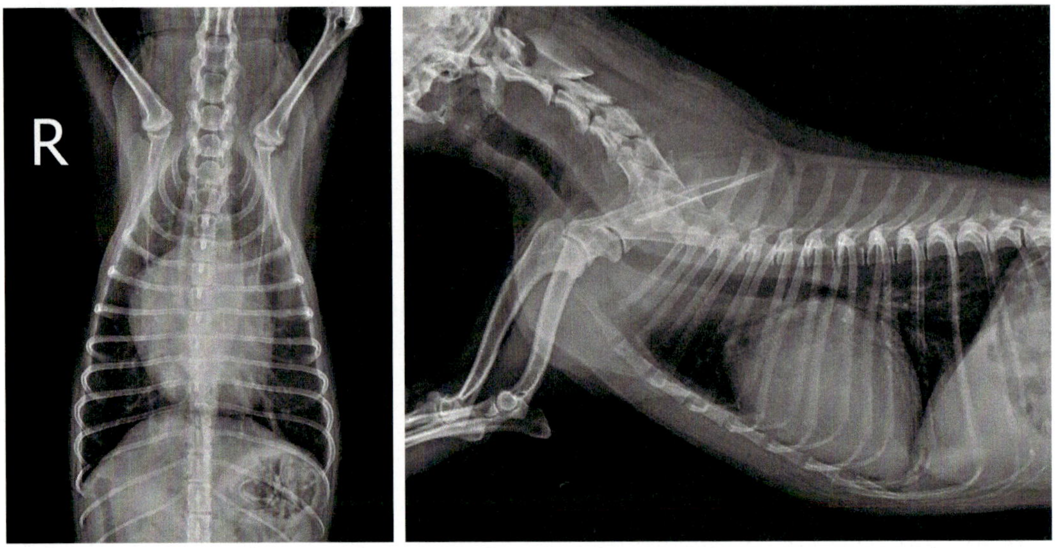

Figure 4-38 ▶ Thoracic radiograph at emergency presentation.

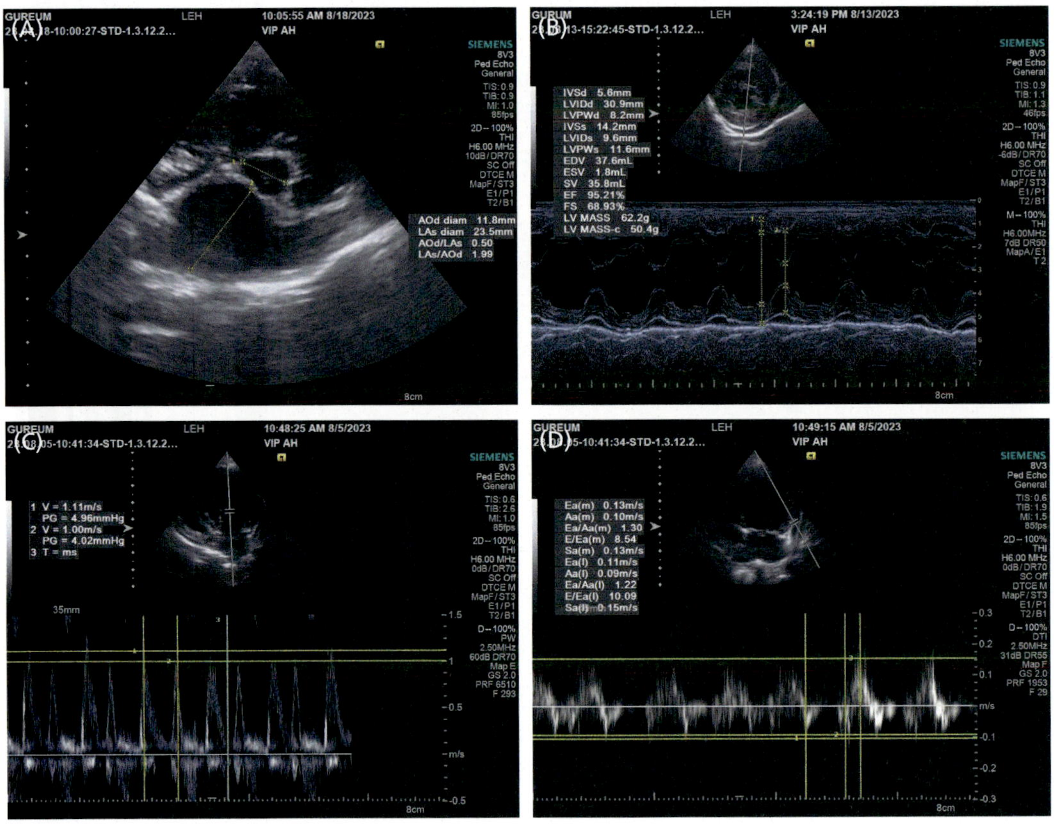

Figure 4-39 ▶ Echocardiographic images at emergency presentation.

Two-dimensional echocardiography confirmed rupture of the chordae tendineae supporting the A1 leaflet of the mitral valve (Figure 4-40). The mitral annulus diameter varied from 11.6-14 mm depending on the beam angle, but all leaflet lengths were greater than 6 mm (Figure 4-40). Based on these findings, the patient's condition was diagnosed as acute heart failure exacerbation due to chordal rupture, leading to mild pericardial effusion. Contrary to the local clinic's diagnosis, the patient did not have concerning pulmonary hypertension, only mild tricuspid valve degeneration.

Treatment and Patient Monitoring

Based on cardiomegaly and left atrial enlargement on thoracic radiographs, atrial and ventricular dilation on echocardiography, signs suggestive of increased left atrial pressure, and mitral regurgitation, the patient was diagnosed with ACVIM C (ISACHC IIIa) stage congestive heart failure and mitral valve regurgitation with mild pericardial effusion. Pericardial effusion was not significant enough to warrant pericardiocentesis. Diuretics were increased (from 0.1 mg/kg q24h to 0.1 mg/kg q12h) to stabilize the patient's condition. Due to the sudden worsening of heart failure caused by partial chordal rupture, TEER with a V-clamp was recommended to the owner.

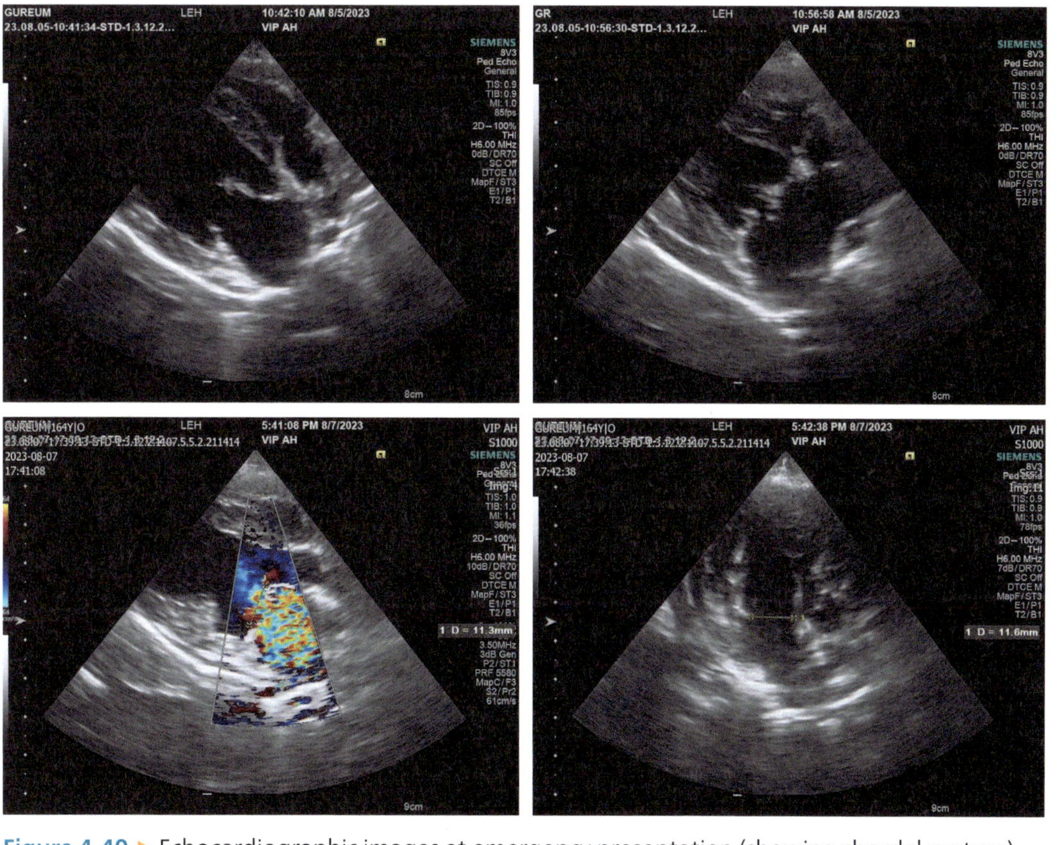

Figure 4-40 ▸ Echocardiographic images at emergency presentation (showing chordal rupture).

The TEER procedure was performed one week later after the patient's condition stabilized. Before the procedure, the lengths of the mitral leaflets and the mitral annulus were measured by echocardiography, and a 14 mm V-clamp (No. 2 V-clamp) was selected. To minimize surgical complications, dalteparin was pretreated from the day before the procedure and continued on the day of the procedure. Lidocaine was administered as a CRI to prevent dangerous ventricular arrhythmias.

A left thoracotomy incision of approximately 8 cm was made to expose the left ventricle. A purse-string suture with 2-0 polypropylene monofilament was placed at the cardiac apex, and a 14F sheath (APT Medical Tech, Hunan, China) was inserted into the left ventricle. The ValveClamp mitral valve repair system, pre-mounted in the sheath, was advanced into the left atrium. The clamp was positioned at the mitral annulus to grasp the valve leaflets as if stapling them (Figure 4-41). Transesophageal echocardiography confirmed a reduction in regurgitant flow. The clamp was then detached from the delivery system, the sheath insertion site was ligated, and the chest wall was sutured to complete the surgery.

Immediately post-procedure, the patient received fentanyl for analgesia and cefotaxime and marbofloxacin for secondary infection prevention. The patient's existing cardiac medications were maintained until recovery.

The day after the procedure, the patient was stable enough to move independently, although intermittent arrhythmias, such as idioventricular rhythm, were observed. Thoracic radiographs showed no pulmonary infiltration, and the heart size had decreased compared to pre-surgery. The V-clamp shadow remained in the same position as immediately after the

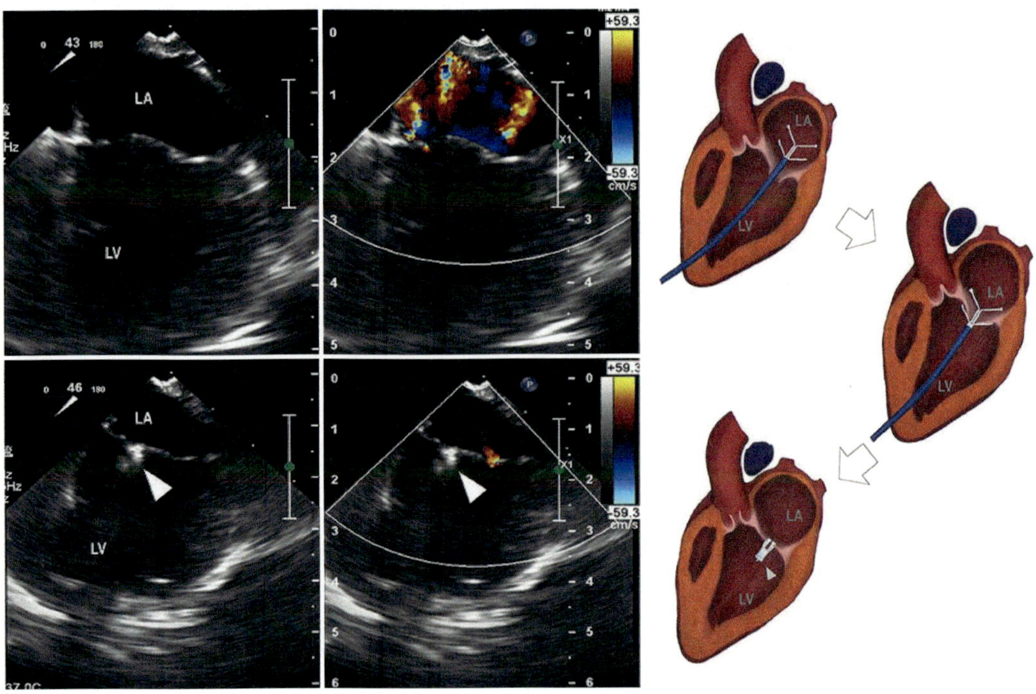

Figure 4-41 ▶ Schematic diagram of the V-clamp TEER procedure performed in this patient.

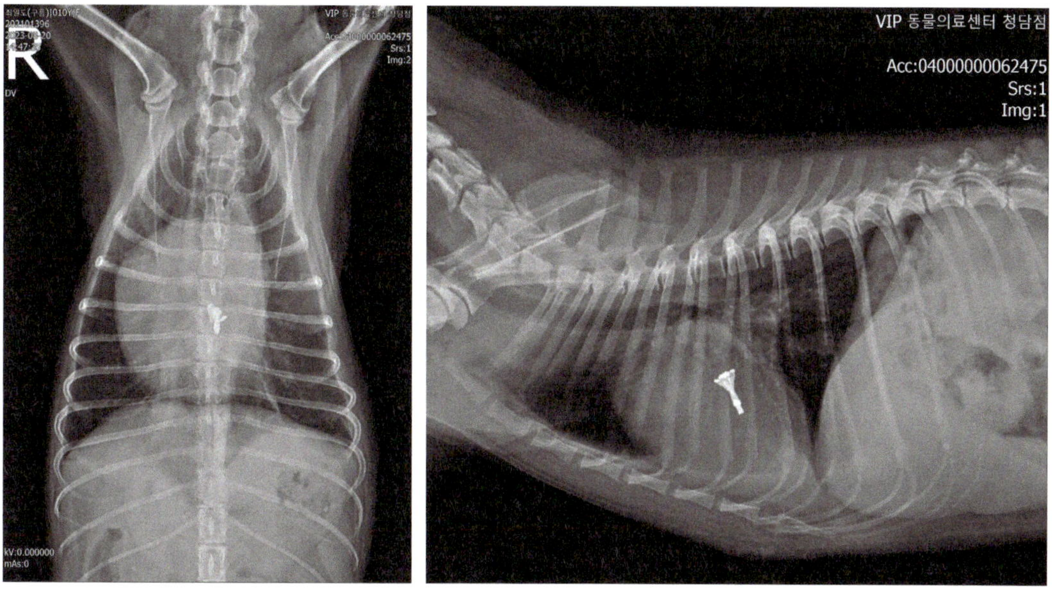

Figure 4-42 ▸ Thoracic radiograph the day after the procedure.

procedure (Figure 4-42). Echocardiography also showed a clear reduction in both left atrial and left ventricular size, with LA/Ao SX measured at 1.54 and LVIDd at 26.7 mm, LVIDd/Ao at 2.3 (Figure 4-43 A&B). Pulsed-wave Doppler showed a mitral inflow E-peak velocity of 0.63 m/sec (ISACHC II), and an extension of the pressure half-time (PHT) and transmitral gradient, indicative of mitral stenosis, were confirmed (Figure 4-43D). Color Doppler also showed a significant reduction in mosaic flow regurgitating into the left atrium (Figure 4-43C).

The patient was discharged with only pimobendan and enalapril, without diuretics, and clopidogrel (once daily) as an antithrombotic agent. The patient re-presented 3 months post-procedure and was reportedly doing well with no particular clinical signs. Thoracic radiographs showed further reduction in heart size, reappearance of the tracheal ventral bend, and no significant change in the clamp's position compared to immediately after the procedure (Figure 4-44).

Discussion

This case demonstrates that TEER using a V-Clamp can be successfully performed in dogs and can be an effective clinical treatment for dogs suffering from mitral valve disease. Contrary to Professor Orton's team's assertion, it was applicable even in dogs weighing less than 5 kg and in cases with mild chordal rupture. Post-procedure, a significant reduction in left atrial and left ventricular size was observed, as evidenced by changes in LA/AO and LVIDd/AO. A slight increase in transvalvular diastolic pressure gradients was noted, but it was not at a level to suspect significant mitral stenosis.

This procedure is functionally similar to human MitraClip or expanded polytetrafluo-

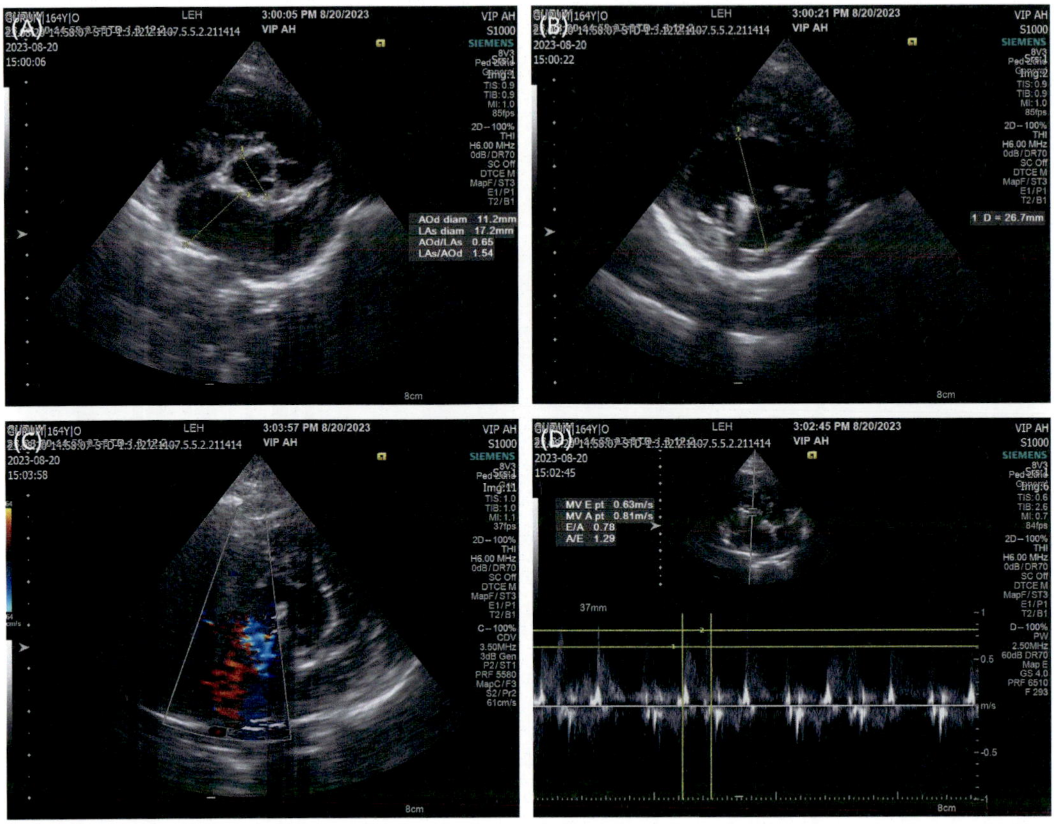

Figure 4-43 ▶ Echocardiographic images the day after the procedure.

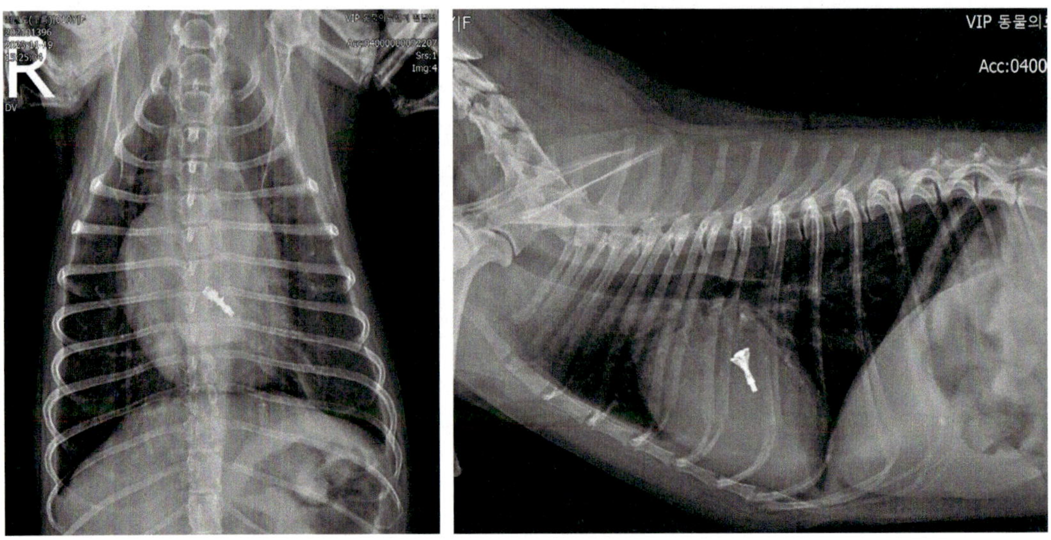

Figure 4-44 ▶ Thoracic radiograph 90 days after the procedure.

roethylene (ePTFE) chordal implantation and does not require cardiopulmonary bypass, which is necessary for open-heart procedures like mitral valvuloplasty. ePTFE chordal implantation involves securing ePTFE artificial chords to the damaged valve and fixing them to the epicardium using ePTFE pledgets. Small studies in dogs using this method have reported high intraoperative mortality and postoperative complications. In contrast, the TEER procedure using the V-Clamp device is known to be technically simpler and associated with fewer intraoperative complications. One experimental study reported a 100% survival rate and rapid recovery, and the Colorado State University team also reported a high surgical success rate (26/28). Surgical failure occurred in 1/28 cases, and postoperative deaths occurred in 2/28 dogs, with one dying 4 hours post-procedure and another 72 hours later. The Colorado State University team's patient selection criteria included severe MR with MR fraction over 50%, moderate to severe left atrial (LA/Ao $\geq$ 2.1) and left ventricular dilation (LVIDdN $\geq$ 1.9), and mitral inflow velocity $\geq$ 1.0 m/s. So far, the 2 year survival rate was reported as 67% by the Colorado University team.

Key factors determining the success rate of this procedure appear to be the patient's heart failure stage and the degree of valve degeneration. To date, there are no reported successful cases in ACVIM D (ISACHC IIIB) patients, with higher success rates observed primarily in ACVIM B2 or C stage patients. The current patient selection criteria proposed by the manufacturer for suitable patients include: 1) valve thickness $\leq$ 3 mm, 2) mitral annulus diameter $\geq$ 13 mm, 3) length of both leaflets $\geq$ 6 mm, 4) LA/Ao $\geq$ 1.6, 5) a gap of $\leq$ 2 mm when both leaflets are closed, 6) no mitral stenosis (PG $\leq$ 2 mmHg, MVA $\geq$ 2 cm^2), 7) no other lesions on the valve such as perforation or calcification, 8) no pathological aortic regurgitation or tricuspid regurgitation, 9) age under 12 years, and 10) no other serious heart diseases on ECG or echocardiography.

Successful performance of this procedure requires 3D transesophageal echocardiography and a C-arm. The core of this procedure is the precise vertical positioning of the clamp in the center of the mitral annulus to grasp both leaflets as if stapling them. Accurately centering the clamp with 2D echocardiography is challenging, making 3D transesophageal echocardiography essential. Additionally, repeatedly attempting to staple can damage the valve. Therefore, it is crucial to properly grasp the mitral leaflets in 1-2 attempts without multiple repetitions.

Since this case, the procedure has been attempted in several other patients. The success of the procedure depends not only on the procedure itself but also on preventing complications like arrhythmias or thrombosis for about a month until the patient stabilizes, and a gradual reduction of cardiac medications as heart size decreases is also very important. It is recommended to administer heparin (50U/kg) pre- and post-procedure along with clopidogrel for 3 months. In another case, a patient suddenly developed symptoms of cerebral infarction one week post-procedure and recovered with additional prescriptions of dalteparin and cerebral blood flow improvers like pentoxifylline, with rivaroxaban subsequently co-administered. Chest tubes can be removed 4 hours post-procedure if no hemorrhage is observed, but in most patients, mild hemorrhagic effusion persists, so they are typically removed 1-2 days later. However, in this particular case, the chest tube was removed the next day. Arrhythmias such as the previously mentioned idioventricular rhythm persisted immediately post-procedure and were typically controlled with lidocaine CRI for 1-2 days. In another case, sudden vagal bradycardia occurred post-procedure, leading to nocturnal syncope, which was controlled by administering glycopyrrolate at night. This patient did not exhibit such findings. According to presentations by the Colorado State University team, severe atrial fibrillation has occurred post-procedure, and such patients could be stabilized with sotalol treatment. In this patient's case, diuretics were discontinued, and cardiac medications were eventually stopped after 12 weeks of monitoring based on heart size at 2, 4, 8, and 12 weeks post-procedure. Due to the reduction in left atrial size after the procedure, diuretic treatment was no longer necessary for heart failure management and was discontinued. After 12 weeks, all cardiac medications were stopped. In other patients, where heart size did not sufficiently decrease, diuretics were gradually reduced over 2-4 week intervals until complete discontinuation, while some patients are still prescribed low-dose diuretics. However, all patients received pimobendan monotherapy or a combination of pimobendan and enalapril for the first two months post-procedure.

In summary, TEER using a V-Clamp is an innovative treatment that can reduce intra-atrial regurgitation in canine mitral regurgitation and holds great potential as a new alternative to open-heart surgery for these patients. However, more data needs to be collected to establish long-term safety and efficacy. Additionally, proper patient selection and meticulous post-procedure patient monitoring and management are crucial for the success of the procedure.

Mitral V-clamp

Medical management primarily treats mitral regurgitation (MR) in dogs, focusing on heart failure. However, mitral valvuloplasty has recently been performed in Japan and Korea. Open-heart surgery, while effective, requires a skilled medical team, cardiopulmonary bypass equipment, and incurs high costs and prolonged hospitalization, limiting its widespread adoption. In humans, non-surgical valve repair (transcatheter edge-to-edge mitral valve repair; TEER) has been developed and is in clinical use, with successful procedures using the MitraClip performed at Asan Medical Center in Korea.

In 2020, Hongyu Medical in China reported the first case of TEER using a V-Clamp in a dog. Subsequently, in 2022 and 2023, Professor Orton's team at Colorado State University presented and discussed cases of MR treatment using the V-clamp at the ACVIM forum. This minimally invasive procedure involves accessing the mitral valve within the heart and using a device called a V-Clamp to staple the mitral valve leaflets together, significantly reducing the amount of mitral regurgitation. Compared to open-heart surgery, TEER offers advantages such as a lower risk of "table death," shorter surgical time, and faster recovery.

However, this procedure was initially considered unsuitable for the Korean context, where small dog breeds (under 5 kg) are prevalent, as it was thought difficult to apply in small dogs or in patients with severely damaged valve leaflets or chordal rupture. This case, however, describes the successful treatment of a small dog (under 5 kg) with acute heart failure exacerbation due to sudden chordal rupture using the V-Clamp.

CHAPTER 05

New Medical and Surgical Therapeutics

5.1 New Medical Therapeutics and Approaches

New Medications and New Findings

TEST Study (Torasemide vs. Furosemide in Dogs with DMVD)

Background: Furosemide is the recommended loop diuretic for congestive heart failure (CHF) secondary to degenerative mitral valve disease (DMVD) in dogs, though limitations include short duration of action and diuretic resistance. Torasemide, a more potent loop diuretic, may provide superior efficacy.

Methods: In two prospective, randomized, single-blinded trials, 366 dogs with CHF due to DMVD received either torasemide once daily or furosemide twice daily for three months, in addition to standard CHF therapy. Primary outcome was treatment success; secondary outcome was time to composite cardiac endpoint.

Results: Torasemide was noninferior to furosemide for short-term treatment success. Importantly, torasemide significantly reduced the risk of cardiac death, euthanasia, or CHF worsening, with an acceptable safety profile.

Conclusions: Torasemide is an effective once-daily alternative to furosemide in dogs with DMVD, offering comparable clinical control and potential prognostic benefits.

Reference:

Chetboul, V., Pouchelon, J.-L., Menard, J., Blanc, J., Desquilbet, L., Petit, A., Rougier, S., Lucats, L. & Woehrle, F.; TEST Study Investigators. (2017). Short-Term Efficacy and Safety of Torasemide and Furosemide in 366 Dogs with Degenerative Mitral Valve Disease: The TEST Study. J Vet Intern Med, 31, 1629-1642.

Spironolactone

This retrospective study aimed to determine if adding spironolactone to the standard treatment for dogs with congestive heart failure (CHF) caused by severe degenerative mitral valve disease (DMVD) would improve their outcome. The standard therapy included furosemide, enalapril, and pimobendan.

The study analyzed data from 81 dogs and divided them into two groups: Group S, which received spironolactone in addition to the standard therapy, and Group C, which received only the standard therapy. The researchers compared a composite primary endpoint of all-cause mortality, cardiac-related death, or euthanasia due to worsening CHF.

The key findings were:

- The primary endpoint occurred in
- **48% of dogs in Group S and 60% of dogs in Group C**, but this difference was **not statistically significant**.
- There were also no significant differences in all-cause mortality or cardiac-related death/euthanasia between the two groups.
- Spironolactone was **well-tolerated**, and no adverse events related to the medication were observed.

In conclusion, the study found that adding spironolactone to the standard treatment regimen did not provide a significant benefit for dogs with severe DMVD and CHF. The authors suggested that limitations of the study, such as its retrospective design and small sample size, may have influenced the results.

Reference:

Stepien RL, Hallowell MG. Efficacy of spironolactone in dogs with congestive heart failure attributable to severe degenerative mitral valve disease. J Vet Intern Med. 2014 Nov-Dec;28(6):1756-63. doi: 10.1111/jvim.12450. Epub 2014 Sep 29. PMID: 25265439.

Hydrochlorothiazide

Overview:

This retrospective study analyzed the short-term and long-term effects of hydrochlorothiazide (HCTZ) in 38 dogs with relapsing congestive heart failure (CHF) due to myxomatous mitral valve disease (MMVD) who had shown resistance to other diuretics. The study evaluated changes in clinical outcomes, laboratory parameters, and echocardiographic measurements.

Key Findings:

- **Long-Term Efficacy:** HCTZ was effective for the long-term management of recurrent CHF, allowing for a reduction in or discontinuation of other diuretics. This suggests HCTZ is a valuable option for managing diuretic resistance and maintaining clinical stability.

- **Short-Term Effects:** In the short term, HCTZ was associated with an increase in creatinine and urea levels, indicating a potential impact on renal function. Therefore, **close monitoring of kidney parameters** is crucial, especially during the initial phase of treatment.
- **Conclusion:** HCTZ can be a successful long-term therapeutic option for dogs with diuretic-resistant CHF. However, vigilance regarding potential adverse effects on renal function is essential for proper patient management.

Reference:

Romito, G. et al. (2025). Short- and long-term effects of hydrochlorothiazide in dogs with relapsing congestive heart failure due to myxomatous mitral valve disease: a retrospective analysis of 38 cases. ***J Vet Cardiol****, 60, 57-69.*

Angiotensin-converting enzyme inhibitors

Overview:

This systematic review and meta-analysis evaluated the efficacy and safety of angiotensin-converting enzyme inhibitors (ACEIs) in dogs with preclinical myxomatous mitral valve disease (MMVD). It synthesized data from multiple studies to assess the impact of ACEIs on preventing CHF and reducing mortality.

Key Findings:

- **Dogs with Cardiomegaly (ACVIM Stage B2):** The analysis showed that ACEI administration provided **no significant benefit** in preventing CHF or reducing mortality. This finding supports current clinical guidelines, which do not recommend routine preventative ACEI therapy in this stage.
- **Dogs without Cardiomegaly (ACVIM Stage B1):** The evidence suggested a possible reduction in the risk of CHF, but the **certainty of this evidence was very low**.
- **Conclusion:** This study highlights the lack of strong evidence to support the preventative use of ACEIs in asymptomatic MMVD dogs. It emphasizes the importance of using a stage-based approach to treatment and avoiding unnecessary medication.

Reference:

Donati, P. et al. (2022). Angiotensin-converting enzyme inhibitors in preclinical myxomatous mitral valve disease in dogs: systematic review and meta-analysis. ***J Small Anim Pract****, 63, 306-318.*

Alacepril

Overview:

This study investigated the safety, tolerance, and efficacy of alacepril—a relatively new angiotensin-converting enzyme (ACE) inhibitor—in dogs with mitral valve disease (MVD) that also had a persistent cough. The study was a prospective, multi-center trial that enrolled

42 dogs with echocardiographic or radiographic evidence of cardiac enlargement (ACVIM stage B2). The dogs were treated with oral alacepril (1.0–3.0 mg/kg/day) for at least four weeks.

Key Findings

- Alacepril was generally well-tolerated in most dogs. Only one dog (2.4%) experienced complications such as appetite loss, lethargy, and vomiting, which resolved after discontinuing the drug for a few days and reducing the dose.
- The treatment was effective in suppressing cough in a significant portion of the dogs. Cough either resolved completely or improved in 20 of the 36 dogs (55.6%) that were re-evaluated after four weeks.
- Dogs in the effective group showed a significant decrease in the left ventricular end-diastolic internal diameter (LVIDDN), while the ineffective group saw an increase.
- Higher baseline levels of atrial natriuretic peptide (ANP), N-terminal pro-B-type natriuretic peptide (NT-proBNP), and E wave velocity were found to be correlated with the ineffectiveness of alacepril.

Conclusion

Alacepril is a safe and well-tolerated treatment for dogs with MVD and cough, with over half of the dogs in the study showing improvement or resolution of their cough. The study suggests that the drug's effectiveness might be influenced by a dog's pre-treatment cardiac loading conditions, as indicated by elevated levels of ANP and NT-proBNP. The authors conclude that alacepril is a promising treatment for improving the quality of life in dogs with early-stage MVD. However, they note that a large-scale, placebo-controlled trial is needed to confirm its long-term efficacy.

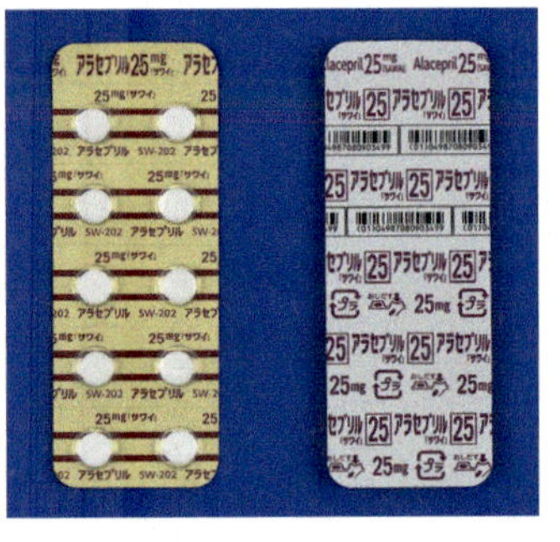

Figure 5-1 ▶ New angiotensin-converting enzyme inhibitor (ACEI), Alacepril

Reference:

Hori, Y. et al. (2018). Effects of the angiotensin-converting enzyme inhibitor alacepril in dogs with mitral valve disease. ***J Vet Med Sci****, 80, 1373–1378*

Sildenafil Study (Pulmonary Hypertension in Dogs)

Background: Pulmonary hypertension in dogs is associated with poor prognosis and limited therapeutic options. Sildenafil, a phosphodiesterase-5 inhibitor, has shown promise in alleviating pulmonary vascular resistance.

Methods: Client-owned dogs with naturally occurring pulmonary hypertension were prospectively treated with oral sildenafil. Clinical outcomes including exercise tolerance, respiratory function, quality of life, and survival were evaluated.
Results: Sildenafil administration resulted in significant improvement in clinical signs and exercise capacity. Adverse events were minimal and well tolerated. Treated dogs demonstrated a longer median survival compared with controls.
Conclusions: Sildenafil is a safe and effective therapeutic option for canine pulmonary hypertension, improving both quality of life and survival.

Reference:

Johnson, L.R. & Stern, J.A. (2020). Clinical features and outcome in 25 dogs with respiratory-associated pulmonary hypertension treated with sildenafil. ***J Vet Intern Med****, 34, 65-73.*

Pulmonary Vasodilators

Overview:

This comparative study investigated the cardiovascular effects of three pulmonary vasodilators—sildenafil, beraprost, and their combination—in canine models with mitral valve disease. The goal was to identify the most effective and safe treatment for managing pulmonary hypertension.

Key Findings:

- **Sildenafil:** While demonstrating a potent vasodilating effect on the pulmonary artery, sildenafil had a negative side effect of increasing left-heart loading.
- **Beraprost:** Beraprost showed a weaker vasodilating effect than sildenafil but did not negatively impact the left heart, suggesting a safer profile in that regard (Figure 5-2).

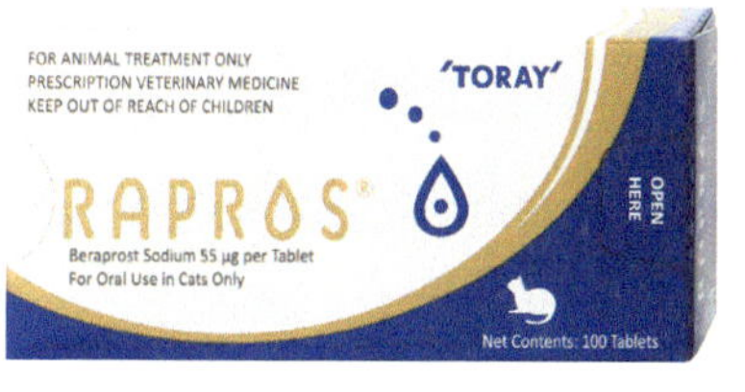

Figure 5-2 ▶ Pulmonary Vasodilator, Beraprost

- **Combination Therapy:** The combination of sildenafil and beraprost produced the most significant vasodilating effects in both pulmonary and systemic circulation **without negatively affecting left-heart loading**.
- **Conclusion:** For severe cases of pulmonary hypertension, the combination of sildenafil and beraprost is a superior treatment option. It leverages the synergistic effects of both drugs to maximize therapeutic benefits while minimizing the risk of adverse cardiac effects.

Reference:

Yuchi, Y. et al. (2024). Comparative Study of Cardiovascular Effects of Selected Pulmonary Vasodilators in Canine Models of Mitral Valve Disease. ***Biology****, 13, 311.*

Sacubitril/valsartan (Entresto)

Overview:

This article details a prospective, randomized, double-blind, placebo-controlled pilot study of **sacubitril/valsartan** (Entresto; Figure 5-3) in dogs with cardiomegaly secondary to myxomatous mitral valve disease (MMVD), corresponding to ACVIM Stage B2. The primary objective was to determine the pharmacodynamic effects of this drug on the renin-angiotensin-aldosterone system (RAAS), natriuretic peptide concentrations, and other cardiovascular parameters.

Key Findings:

- **Pharmacodynamic Effects:** The study observed that sacubitril/valsartan led to **significant pharmacodynamic changes** in the treated dogs. Specifically, the drug caused a rise in plasma renin activity and natriuretic peptide concentrations, while simultaneously decreasing aldosterone concentrations.

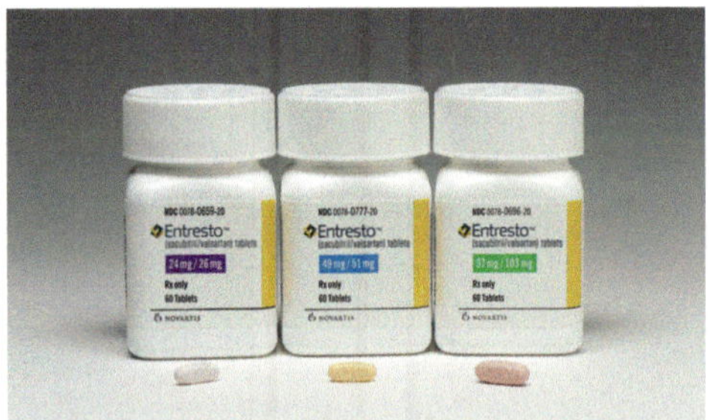

Figure 5-3 ▸ Sacubitril/valsartan (Entresto)

- **Safety:** No significant differences were noted in systemic arterial pressure, tests of renal function, or serum electrolyte concentrations between the treatment and placebo groups. This suggests that the drug was well-tolerated in this cohort.
- **Conclusion:** This pilot study establishes that sacubitril/valsartan has predictable pharmacodynamic effects in dogs with asymptomatic MMVD. While the study did not evaluate long-term clinical outcomes, the findings suggest that the drug is well-tolerated and warrants further investigation as a potential therapeutic agent for canine MMVD.

Reference:

Newhard, D.K. et al. (2018). A prospective, randomized, double-blind, placebo-controlled pilot study of sacubitril/valsartan (Entresto) in dogs with cardiomegaly secondary to myxomatous mitral valve disease. ***J Vet Intern Med****, 32, 1693-1701.*

Tolvaptan

Tolvaptan is a selective, competitive vasopressin V2 receptor antagonist (Figure 5-4). In human medicine, it is used to treat hyponatremia associated with congestive heart failure and the syndrome of inappropriate antidiuretic hormone. It acts as an aquaretic agent, which means it facilitates free water clearance in the kidneys, leading to diuresis without affecting the balance of electrolytes. This makes it a potential alternative to loop diuretics, which can worsen renal function with long-term use. A study on dogs with myxomatous mitral valve disease and concurrent cardiorenal syndrome found that tolvaptan was effective and safe in treating the condition, improving renal function without negatively affecting cardiac or hepatic parameters.

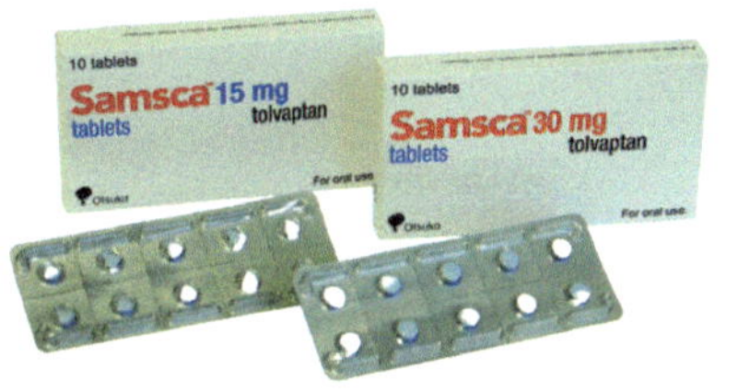

Figure 5-4 ▶ Selective, competitive vasopressin V2 receptor antagonist, Tolvaptan

Background and Objectives: This study evaluated the use of tolvaptan, a selective vasopressin V2 receptor antagonist, in dogs with myxomatous mitral valve disease (MMVD) and concurrent cardiorenal syndrome (CRS). While loop diuretics like torsemide are the primary treatment for congestive heart failure (CHF) in dogs, their long-term use can worsen renal function. Tolvaptan, known in human medicine to be a safe and effective diuretic that does not negatively impact renal function or electrolyte balance, was hypothesized to be a beneficial alternative for dogs with CHF and CRS.

Methods: The study included 41 client-owned dogs with MMVD and CHF (ACVIM Stage C and D) that also had evidence of CRS. The dogs were divided into three groups based on their diuretic treatment:

- Group 1: Torsemide dose was reduced by 25-50%.
- Group 2: 50% of the torsemide dose was replaced with tolvaptan (3.3 mg tolvaptan equivalent to 0.1 mg torsemide).
- Group 3: The entire torsemide dose was replaced with tolvaptan.

Researchers monitored renal markers (serum urea nitrogen, creatinine, and symmetric dimethylarginine [SDMA]), as well as echocardiographic and hepatic parameters, on Day 0, Day 30, and Day 60.

Results: The study found that serum levels of urea nitrogen, creatinine, and SDMA were significantly lower in Group 2 and Group 3 on Day 30 and Day 60 compared to Day 0 ($P < 0.01$). This indicates that tolvaptan was effective in inducing diuresis without worsening renal function. Additionally, serum sodium and potassium levels were improved in these groups. In contrast, Group 1 did not show a significant improvement in these renal markers compared to Day 0. Echocardiographic and hepatic parameters did not change significantly across any of the groups or over time.

Conclusion and Clinical Importance: The study concluded that tolvaptan is an effective and safe treatment for CRS in dogs with advanced MMVD and CHF. The addition or replacement of tolvaptan helped to lessen the severity of CRS without negatively impacting cardiac or hepatic function. The authors noted that this was not a large-scale, double-blinded study, and that a more controlled trial is needed to confirm the optimal dosage and long-term efficacy and safety of tolvaptan in dogs with CHF and CRS.

Sodium-glucose co-transporter 2 (SGLT2) inhibitors

Overview:

This review article explores the potential use of **sodium-glucose co-transporter 2 (SGLT2) inhibitors** for treating canine myxomatous mitral valve disease (MMVD). SGLT2 inhibitors are a class of human anti-diabetic drugs, with some having been approved for treating chronic heart failure in people. The authors emphasize the challenge of finding the right drug and dose for dogs and highlight that non-clinical information from human drug approval packages, where canines are often used, could be invaluable.

Key Findings:

- **Pharmacological Profile:** The review provides detailed pharmacokinetic and pharmacodynamic data on various SGLT2 inhibitors in dogs, including dapagliflozin (Figure 5-5) and empagliflozin. It compares drug exposure and metabolism between dogs and humans, noting that dogs can serve as a non-clinical model for understanding these drugs.

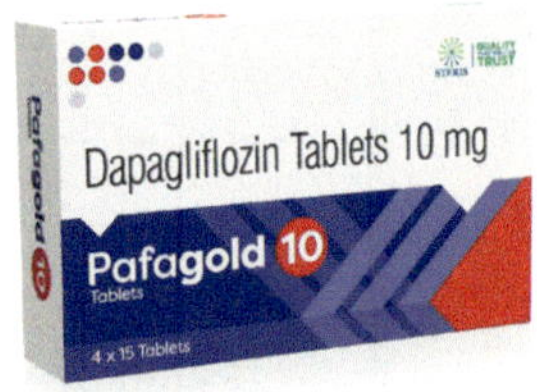

Figure 5-5 ▶ Sodium-glucose co-transporter 2 (SGLT2) inhibitor, dapagliflozin

- **Therapeutic Potential:** The article discusses the potential cardioprotective mechanisms of SGLT2 inhibitors, such as improving cardiac function and reducing inflammation and fibrosis. It suggests that these effects could be beneficial for treating canine MMVD, given the similarities between human and canine heart failure.
- **Conclusion:** SGLT2 inhibitors show promise as a future treatment for canine MMVD. The authors conclude that data from non-clinical studies can help veterinarians select appropriate drugs and dosages, paving the way for further clinical trials in dogs.

Reference:

Umezawa, M. et al. (2025). Sodium-glucose co-transporter 2 inhibitors: Prospects for canine myxomatous mitral valve disease and finding the "right drug" and the "right dose" for dogs. ***J Vet Med Sci****, 87, 647-666.*

Synthetic canine B-type natriuretic peptide (syncBNP)

This study investigated the safety, feasibility, and tolerance of administering synthetic canine B-type natriuretic peptide (syncBNP) to privately owned dogs via subcutaneous (SC) injection. The research was conducted as a Phase 1 dose-escalation trial involving six dogs: four healthy dogs and two dogs with stage B1 mitral valve disease.

Key Findings:

- **Safety and Tolerance:** SyncBNP administered subcutaneously was found to be feasible, safe, and well-tolerated at all tested doses (2.5, 5, and 10 μg/kg). No major adverse events were observed in any of the dogs.
- **Biological Activity:** The 5 μg/kg dose of syncBNP significantly increased the concentration of plasma cGMP (3',5'-cyclic guanosine monophosphate), a secondary messenger for natriuretic peptides, at both 45 and 120 minutes after injection. This increase suggests that the syncBNP successfully bound to and activated natriuretic peptide receptors.
- **Hemodynamic Effects:** The study found no significant changes in systolic blood pressure or heart rate after the 5 μg/kg dose.
- **Implications:** The results suggest that syncBNP could be a viable treatment option for dogs with heart disease. The findings pave the way for future studies to investigate the long-term clinical benefits of both acute and chronic syncBNP administration in dogs with more advanced heart conditions.

This study is the first to report the use of syncBNP in privately owned dogs in a clinical veterinary setting.

5.2 Drug Trials

The BESST Study

Overview:

The **BEnazepril Spironolactone STudy (BESST)** was a prospective, randomized, double-blind, placebo-controlled trial evaluating the clinical efficacy of a fixed-dose combination of benazepril and spironolactone in dogs with congestive heart failure (CHF) due to myxomatous mitral valve disease (MMVD). The study aimed to assess the effects of this combination on a composite endpoint of all-cause mortality, euthanasia due to cardiac failure, and signs of clinical deterioration.

Key Findings:

- **Reduced Endpoint Events:** The benazepril-spironolactone group showed a statistically significant reduction in the composite endpoint compared to the placebo group.
- **Improved Clinical Signs:** The dogs receiving the combination therapy also showed a higher percentage of improvement in clinical signs such as dyspnea and coughing.
- **Conclusion:** The study concluded that the fixed-dose combination of benazepril and spironolactone is an effective and safe treatment for dogs with MMVD-induced CHF, providing significant clinical benefits.

Reference:

Coffman, M. et al. (2021). Clinical efficacy of a benazepril and spironolactone combination in dogs with congestive heart failure due to myxomatous mitral valve disease: The BEnazepril Spironolactone STudy (BESST). ***J Vet Intern Med****, 35, 1673-1687.*

The VALVE Trial

Overview:

This study, known as the **VALVE trial**, investigated the efficacy of adding the ACE inhibitor ramipril to a dual-therapy regimen of furosemide and pimobendan in dogs with degenerative mitral valve disease. The goal was to determine if this "triple therapy" (TT) improved time to a cardiac endpoint (such as death or treatment failure) compared to dual therapy (DT). The study was a randomized, double-blind, placebo-controlled trial.

Key Findings:

- No Significant Benefit: The study found no statistically significant difference in the time to the combined cardiac endpoint between the triple-therapy group and the dual-therapy group.
- Adverse Events: While both groups showed similar adverse event profiles, dogs in the ramipril group experienced a statistically significant increase in serum creatinine levels compared to the placebo group. This suggests that adding ramipril may negatively impact renal function.
- Conclusion: The findings suggest that adding ramipril to a combination of furosemide and pimobendan does not offer a significant clinical benefit in dogs with mitral valve degeneration. The observed increase in creatinine levels warrants caution and reinforces the need for close monitoring of renal parameters when using this triple therapy.

Reference:

Wess, G. et al. (2020). Efficacy of adding ramipril (VAsotop) to the combination of furosemide (Lasix) and pimobendan (VEtmedin) in dogs with mitral valve degeneration: The VALVE trial. ***J Vet Intern Med****, 34, 2232-2244.*

Sentorpil® ForteGold

Overview:

This randomized, double-blind study compared a novel fixed-dose combination tablet (Sentorphil® ForteGold; Figure 5-6) with compounded powdered medications for the treatment of myxomatous mitral valve disease in dogs. The study assessed the clinical effectiveness, safety, and owner compliance for both treatments.

Key Findings:

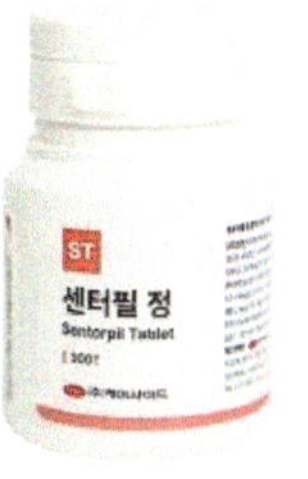

Figure 5-6 ▸ Enalapril, torsemide, spironolactone and pimobendane combination Tablet, Sentorpil ForteGold

- Superior Efficacy: The fixed-dose tablet resulted in **faster, more significant, and sustained clinical improvements** compared to the compounded medication.
- **Safety and Compliance:** The tablet was associated with fewer adverse events and better owner compliance due to its ease of administration and accurate dosing.
- **Conclusion:** This study concludes that the new fixed-dose formulation is clinically superior, safer, and more convenient for owners. Its use leads to better treatment outcomes and is likely to improve the overall management of canine MMVD.

Reference:

Park, J. et al. (2025). Clinical evaluation of a novel fixed-dose combination tablet (Sentorpil ForteGold) versus compounded powdered medications in dogs with myxomatous mitral valve disease: a randomized, double-blind study. ***Front Vet Sci****, 12, 1622383.*

Fixed-dose combination tablet of benazepril and pimobendan (Fortekor Plus)

Overview:

This article describes a randomized, non-inferiority clinical trial conducted in Japan to evaluate a **fixed-dose combination tablet of benazepril and pimobendan** (Fortekor Plus; Figure 5-7) for dogs with congestive heart failure (CHF) due to myxomatous mitral valve disease (MMVD). The study aimed to determine if this fixed-dose combination was as effective as the separate, registered formulations of the two drugs.

Key Findings:

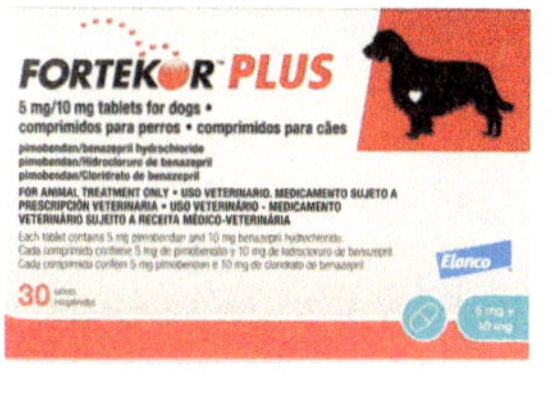

Figure 5-7 ▸ Benazepril and pimobendane combination Tablet, Fortekor Plus

- **Study Design:** The trial was a three-arm, masked study. The test group received the fixed-dose combination tablet twice daily. Two control groups received the individual drugs: one with benazepril and pimobendan both twice daily (Control I), and another with benazepril once daily and pimobendan twice daily (Control II).
- **Primary Outcome:** The primary endpoint was a composite of treatment failure, which included the need for additional medication, euthanasia, or death.
- **Results:** The study's results demonstrated that the fixed-dose combination was

non-inferior to the separate formulations in managing CHF in dogs.

- **Conclusion:** This trial confirms that a fixed-dose combination of benazepril and pimobendan is a valid and effective treatment option for dogs with MMVD-induced CHF. This formulation can simplify treatment, potentially improving owner compliance, without compromising efficacy compared to administering the drugs separately.

Reference:

King, J.N. et al. (2018). Evaluation of a fixed-dose combination of benazepril and pimobendan in dogs with congestive heart failure: a randomized non-inferiority clinical trial. ***J Vet Sci****, 19, 117–128*

Efficacy of Combination therapy

Overview:

This multicenter retrospective study compared the effects of different medical therapies on cardiac death in dogs with myxomatous mitral valve disease (MMVD) at American College of Veterinary Internal Medicine (ACVIM) stage C. The study compared dual therapy (**DT**; furosemide and pimobendan), triple therapy (**TT**; DT plus benazepril), and quadruple therapy (**QT**; TT plus spironolactone).

Key Findings:

- **No Difference in Cardiac Death:** The study found no statistically significant differences in the rate of cardiac death among the DT, TT, and QT groups.
- **Similar Survival Curves:** The Kaplan-Meier cardiac death survival curves were not statistically different in either unadjusted or adjusted analyses.
- **Conclusion:** The study concluded that, for the dogs included in this sample, triple and quadruple therapies did not provide a higher clinically evident cardiac protective effect compared to dual therapy. The findings suggest that DT was not inferior to TT and QT in terms of preventing cardiac death in these patients.

Reference:

Romito, G. et al. (2025). Comparison of the effects of dual, triple, and quadruple medical therapy on cardiac death in a retrospective cohort of dogs with myxomatous mitral valve disease at American College of Veterinary Internal Medicine stage C: is more necessarily better?. ***J Am Vet Med Assoc****, 2025, 1-11.*

5.3 New Surgical Therapeutics and Approaches

New Cardiac Interventions

Left atrial decompression (LAD) is a palliative, minimally invasive interventional procedure designed to reduce the dangerously high pressure in the left atrium (LA) of the heart. This condition, known as left atrial hypertension, is a direct consequence of severe mitral valve regurgitation in dogs with advanced MMVD. In a healthy heart, blood flows from the left atrium into the left ventricle (LV) and then is pumped out to the body. When the mitral valve degenerates and leaks, a portion of the blood flows backward into the LA with each heartbeat, causing the LA to enlarge and the pressure to build. This elevated pressure is transmitted to the pulmonary veins and capillaries, leading to fluid extravasation into the lungs (pulmonary edema), which is the hallmark of left-sided congestive heart failure (CHF).

The goal of LAD is not to fix the underlying valve problem, but to relieve the pressure in the left atrium by creating a controlled, intentional hole (an "iatrogenic atrial septal defect," iASD) in the muscular wall that separates the left and right atria. This allows blood to shunt from the high-pressure left atrium to the lower-pressure right atrium, thereby "decompressing" the left side of the heart.

The Procedure

The procedure is performed via cardiac catheterization, which is a key aspect of its minimally invasive nature. Here's a step-by-step overview of the technique:

1. Preparation: The dog is placed under general anesthesia.
2. Access: A catheter is inserted into a large vein, typically the jugular vein in the neck.
3. Guidance: Using advanced imaging techniques such as fluoroscopy (continuous X-ray images), transesophageal echocardiography (TEE), and in some cases, intracardiac echocardiography (ICE), the catheter is carefully guided to the right atrium.
4. Puncture: A specialized needle (Figure 5-8) is used to puncture the interatrial septum, the wall separating the right and left atria. This is a critical step, and the imaging guidance is essential to ensure the puncture is made at the thinnest and safest part of the septum (the fossa ovalis).
5. Creation of iASD: After the puncture, a balloon catheter is often used to dilate the newly created hole, ensuring it is large enough to allow for effective shunting of blood (Figure 5-9).

Scientific and Clinical Rationale

The scientific basis for LAD is to immediately and significantly reduce left atrial pressure (LAP), which directly alleviates the signs of CHF. Clinical studies have shown an immediate and substantial drop in LAP after the procedure, with a median decrease of 6 mmHg in one study. This pressure reduction leads to a rapid improvement in clinical signs such as difficulty breathing and coughing, and can allow for a reduction in the dosage of diuretics, which can have beneficial effects on renal function.

Outcomes and Limitations

LAD is considered a palliative treatment, meaning it is aimed at improving the quality of life and controlling symptoms rather than curing the disease. It is often a treatment of last

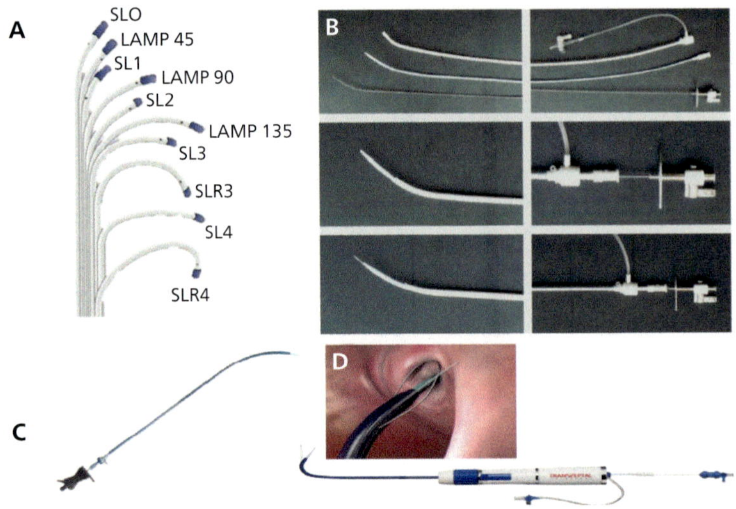

Figure 5-8 ▸ Devices used for trans-septal puncture. Swartz SL transseptal sheaths (A); Abbott transseptal kit (B); NRG RF transseptal needle (C); TSP Crosser kit (D).

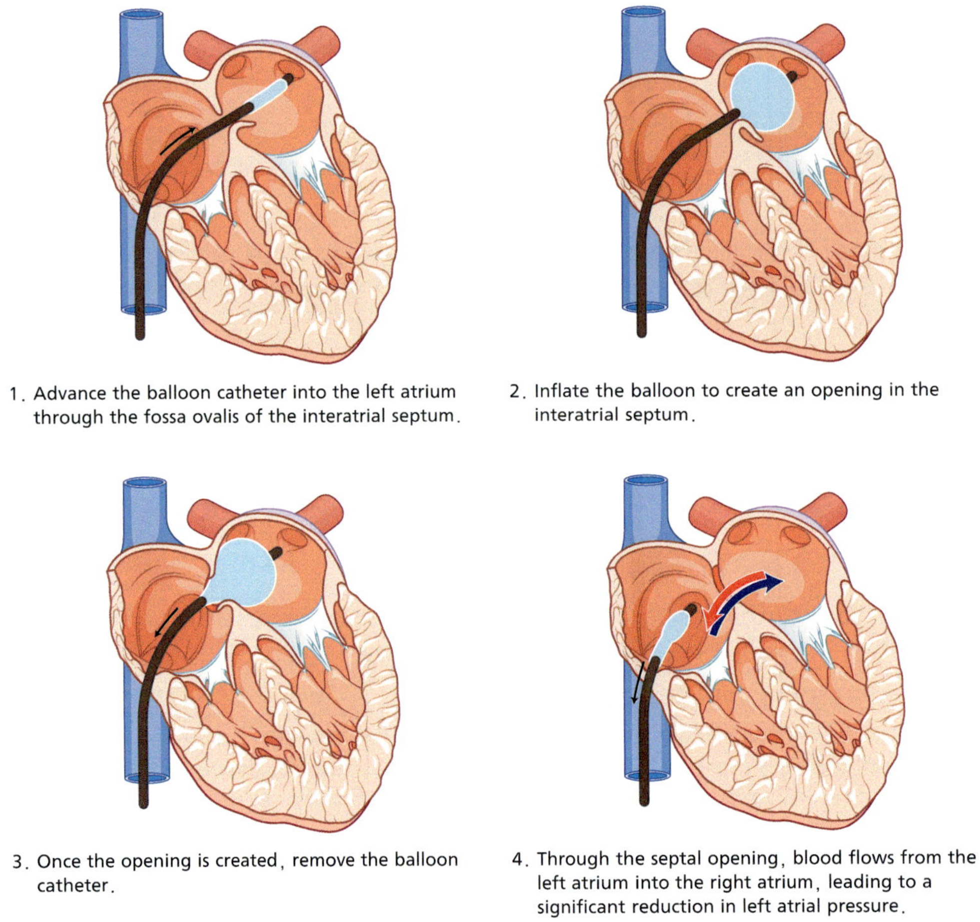

Figure 5-9 ▸ Procedure of left atrial decompression with trans-septal puncture

resort for dogs with advanced (ACVIM Stage C and D) MMVD who are refractory to optimal medical management.

Key outcomes and considerations include:

- Rapid Symptom Relief: Dogs often show immediate improvement in breathing and energy levels.
- Reduced Pulmonary Congestion: By lowering LAP, LAD reduces the fluid buildup in the lungs.
- Median Survival: The median survival time for dogs post-LAD has been reported to be approximately 195 days. This is a significant improvement over the expected outcome for dogs with severe, medically refractory CHF.
- Complications: While minimally invasive, the procedure is not without risks, including potential for pericardial effusion leading to cardiac tamponade, atrial thrombus formation, and arrhythmias. The success of the procedure is highly dependent on the skill of the veterinary cardiologist and the use of precise imaging guidance.
- Complementary to Other Therapies: LAD can be an alternative or a bridge to more definitive surgical treatments like mitral valve repair, which are currently limited by availability and cost. It is also often used in combination with a modified medical therapy plan post-procedure.

Allen JW, Phipps KL, Llamas AA, Barrett KA. Left atrial decompression as a palliative minimally invasive treatment for congestive heart failure caused by myxomatous mitral valve disease in dogs: 17 cases (2018-2019). J Am Vet Med Assoc. 2021 Mar 15;258(6):638-647.

Purpose of the procedure

Designed as a palliative catheter-based method to reduce left atrial pressure and alleviate congestive heart failure (CHF) due to severe myxomatous mitral valve disease (MMVD).

Serves as an alternative when medical therapy is insufficient and surgical mitral repair is not available.

Procedure overview

Approach: Percutaneous, catheter-based transatrial septal puncture to create a communication between the left atrium and right atrium.

Goal: Provide a controlled "pop-off valve" mechanism to decompress the left atrium by shunting blood into the right atrium.

Step-by-step technique

1. **Anesthesia & monitoring:**
    - Dogs placed under general anesthesia.
    - Standard cardiopulmonary monitoring with continuous echocardiographic and fluoroscopic guidance.

2. **Vascular access:**
   - Access obtained via the right femoral vein or jugular vein.
   - A sheath was advanced into the right atrium under fluoroscopic and echocardiographic visualization.

3. **Transseptal puncture:**
   - A specialized transseptal needle and sheath system advanced to the interatrial septum.
   - Puncture performed at the level of the fossa ovalis, guided by fluoroscopy and transesophageal echocardiography (TEE).

4. **Creation of atrial communication:**
   - After puncture, the tract was balloon dilated to enlarge the interatrial communication.
   - Some cases used stent placement to maintain patency.

5. **Confirmation:**
   - Echocardiography confirmed left–to–right shunting of blood.
   - Hemodynamics demonstrated reduced left atrial pressure.

Post-procedure management

Dogs recovered with supportive care, diuretic therapy, and follow-up echocardiography.
Clinical outcome: Significant reduction in pulmonary edema and CHF signs, though considered palliative (not curative), with variable long-term benefit.

Allen J, Phipps K, Barrett K. Transseptal puncture in the dog utilizing three-dimensional transesophageal echocardiographic guidance. J Vet Cardiol. 2024 Feb;51:64-71.

Purpose of the procedure

To describe and evaluate the use of 3D TEE (transesophageal echocardiography) for real–time guidance of transseptal puncture in dogs. Critical for interventions such as left atrial decompression or structural cardiac interventions (e.g., experimental mitral valve repair devices).

Procedure overview

Technique: Transseptal puncture performed under direct 3D TEE visualization, which provides detailed spatial orientation of the interatrial septum.
Advantage: Increased accuracy and safety compared to fluoroscopy or 2D TEE alone.

Step-by-step technique

1. **Patient preparation:**
   - Dogs anesthetized and intubated.
   - Positioned in right lateral recumbency.
   - Continuous ECG and invasive blood pressure monitoring.

2. **TEE probe placement:**
   - A specialized 3D–capable TEE probe inserted into the esophagus.
   - Imaging optimized to visualize the fossa ovalis region of the interatrial septum.

3. **Vascular access & catheter positioning:**
    - Venous access via femoral or jugular vein.
    - A transseptal sheath and dilator advanced into the right atrium under fluoroscopic and echocardiographic guidance.

4. **Septal puncture with 3D TEE guidance:**
    - The transseptal needle positioned against the fossa ovalis.
    - 3D TEE provided real–time en face view of the septum, allowing precise selection of the puncture site.
    - Once confirmed, the needle advanced into the left atrium.

5. **Verification of successful puncture:**
    - 3D TEE confirmed the location, trajectory, and bubble contrast jet indicating left atrial entry.
    - Catheter advanced safely into the LA.

Clinical significance

Demonstrated that 3D TEE improves accuracy, visualization, and operator confidence, while potentially reducing complications (e.g., inadvertent puncture of the aortic root or free wall). Technique establishes a platform for future minimally invasive mitral valve therapies in dogs.

Foulex, P. et al. (2025). A Novel Technique of Left Atrial Decompression Using Intracardiac Echocardiography Guidance in 2 Dogs With Advanced Degenerative Mitral Valve Disease. J Vet Intern Med, 39, e70010.

Overview:

This case report describes a novel technique for **left atrial decompression** (LAD) using intracardiac echocardiography (ICE) guidance in two dogs with advanced degenerative mitral valve disease. This procedure, involving a transseptal puncture followed by balloon atrial septostomy, is presented as a palliative, minimally invasive treatment to reduce elevated left atrial pressure.

Key Findings:

- **Feasibility and Safety:** The procedure was deemed feasible and safe in both dogs, who suffered from severe pulmonary hypertension.
- **Clinical Benefits:** Both dogs showed immediate improvement in clinical signs post-procedure, which included a reduction in left atrial pressure and a decrease in the severity of pulmonary hypertension. One dog's condition improved enough to stop continuous oxygen therapy.
- **Conclusion:** This case report suggests that LAD guided by ICE is a promising, minimally invasive palliative option for dogs with advanced MMVD and severe pulmonary hypertension that are not responsive to conventional medical therapy.

New Hybrid Therapies

Transcatheter Edge-to-Edge Mitral Valve Repair (TEER) is a minimally invasive cardiac procedure for treating severe mitral regurgitation (MR) in dogs with myxomatous mitral valve degeneration (MMVD). Unlike traditional open-heart surgery, which requires a heart-lung bypass machine, TEER is performed on a beating heart, significantly reducing surgical time and invasiveness. This technique is adapted from a similar procedure used in humans, leveraging a veterinary-specific device known as the V-Clamp. The goal is to correct the regurgitant blood flow by creating a "double orifice" in the valve, thereby lessening the strain on the left side of the heart and improving clinical signs of heart failure.

The Procedure: Mechanism and Technique

The TEER procedure involves a hybrid approach, combining a small surgical incision with catheter-based intervention. It is typically performed under general anesthesia and guided by advanced imaging, including transesophageal echocardiography (TEE) and fluoroscopy, which provide real-time, three-dimensional visualization of the heart.

The steps are as follows:

1. Access: A small thoracotomy (incision into the chest wall) is made, usually at the fifth or sixth intercostal space. A catheter is then inserted through the cardiac apex (the tip of the heart; Figure 5-10).
2. Device Delivery: The V-Clamp device, which is made of a Nitinol alloy, is delivered via the catheter into the left ventricle and positioned across the mitral valve.
3. Leaflet Grasping: The device is carefully manipulated to grasp the edges of the anterior and posterior mitral valve leaflets, most commonly at the central segments (A2 and P2) where prolapse or flail is most common (Figure 5-11).
4. Implantation: Once the correct position is confirmed, the V-Clamp is deployed, suturing or "clamping" the leaflets together to form a central bridge. This reduces the large, single regurgitant orifice into two smaller, functioning orifices, effectively decreasing the backward flow of blood.
5. Post-Procedure: The catheter is withdrawn, and the incision is closed. The clamp remains as a permanent implant. The average procedural time is 60-80 minutes.

Patient Selection and Clinical Outcomes (Table 5-1)

Not all dogs with MMVD are candidates for TEER. Patient selection is critical for success. The ideal candidates are small-breed dogs (over 3 kg), typically in the late ACVIM Stage B2 or Stage C of the disease, with an appropriate mitral valve anatomy (e.g., central leaflet prolapse without a cleft).

Clinical studies and case series have shown promising results:

- Procedural Success: The procedural feasibility rate is high, with one study reporting a 96% success rate for device delivery.

CENTRAL ILLUSTRATION: Beating-Heart Mitral Valve Repair Procedure

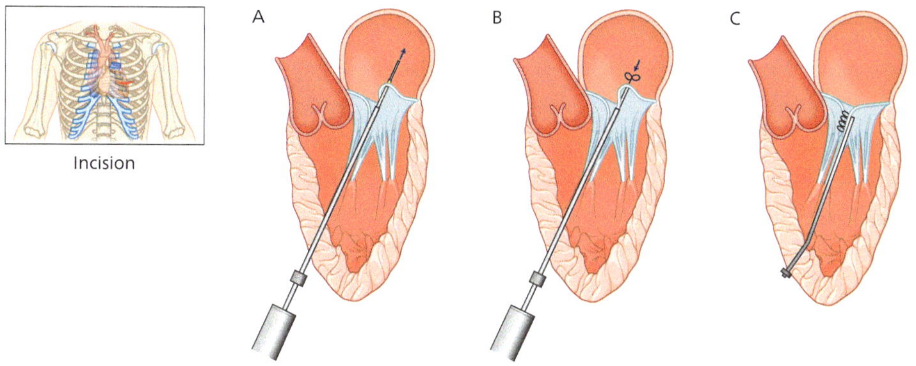

Gammie, J.S. al. J Am Coll Cardiol. 2018;71(1):25-36.

Step A: In the first step, a specialized surgical tool, a mitral valve guide, is inserted into the heart's left ventricle, near the mitral valve. The guide is used to align and position the repair device. The mitral valve itself consists of two leaflets and a network of chordae tendineae, which are thin, tendon-like cords that anchor the leaflets to the heart muscle.

Step B: In the second step, the guide is used to steer a suture or clip through the two leaflets of the mitral valve. This is the core of the repair. The goal is to bring the two leaflets together at the point where they are leaking, thereby restoring proper valve function. The leaflets are not cut; they are gently brought into alignment and secured.

Step C: In the final step, the suture or clip is cinched down, effectively creating a double-orifice mitral valve. This new shape allows the valve to close properly, preventing blood from leaking back into the left atrium (a condition known as mitral regurgitation). The surgical tool is then withdrawn, and the heart's normal pumping action is restored with the repaired valve.

Figure 5-10 ▶ Procedure of left atrial decompression with trans-septal puncture

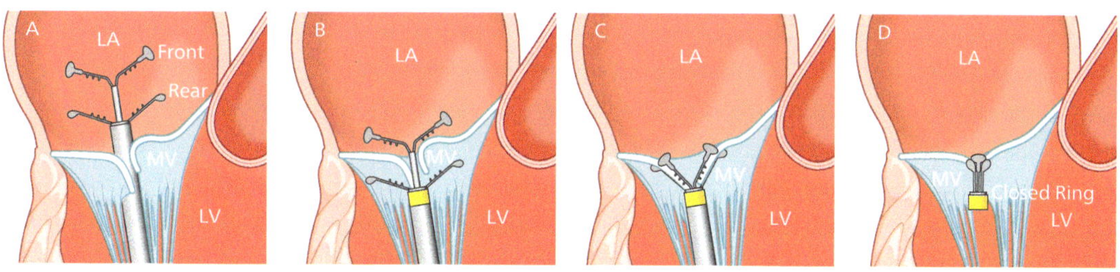

Leaflet capture and clamping (A-D) The device arms are opened and aligned across the mitral valve. Both anterior and posterior mitral leaflets are grasped, creating a double-orifice configuration (similar to the human MitraClip technique). Echocardiography confirms leaflet capture, secure positioning, and reduction of mitral regurgitation. If satisfactory, the device is released from the delivery catheter. The clamp reduces the regurgitant orifice, lowers left atrial pressure, and alleviates congestive heart failure symptoms.

Figure 5-11 ▶ Procedure of mitral V-clamp (transcatheter edge-to-edge repair, TEER). The V-clamp delivery system is advanced through the septal puncture into the left atrium. Under real-time transesophageal echocardiography, the clamp is positioned above the mitral valve leaflets at the regurgitant jet origin.

Table 5-1 ▶ Patient Selection Guidelines for Canine TEER (Mitral V-clamp)

Category	Guideline Criteria	Rationale / Notes
Underlying disease	Confirmed myxomatous mitral valve disease (MMVD) with severe mitral regurgitation (MR).	TEER is designed for primary MR due to leaflet malcoaptation. Not indicated for congenital or secondary MR (e.g., DCM, endocarditis).
Clinical stage (ACVIM)	Symptomatic dogs: Stage C (CHF due to MR) or Stage D (refractory CHF). Selected Stage B2 (preclinical with marked remodeling) may be considered if rapid progression.	Aligns with cases where medical therapy alone is insufficient or prognosis is poor.
Medical therapy optimization	Dog must be on maximal tolerated medical therapy (pimobendan, ACE−inhibitor/ARB, diuretics, ± spironolactone) with persistent CHF signs.	TEER is considered adjunctive or rescue therapy after failure of medical stabilization.
Echocardiographic MR severity	– Central jet with severe MR (vena contracta >0.5 cm, or regurgitant fraction >50%, or E−wave >1.5 m/s). – Anatomy suitable for edge−to−edge repair (sufficient leaflet tissue, not massively calcified).	Successful TEER requires adequate grasping of valve leaflets and effective MR reduction.
Cardiac remodeling	– LA/Ao typically >2.0 – LVIDDn >1.7 – Preserved or supranormal LV systolic function (FS% >25%).	Ensures MR severity is clinically relevant and LV contractility remains adequate.
Body weight & size	Medium to large breeds (>5 − 6 kg) preferred. Device delivery and manipulation are technically difficult in very small dogs.	Current catheter sizes and clamp systems are not optimized for <5 kg patients.
Exclusion criteria	– Active endocarditis – Severe pulmonary hypertension with RV failure – Severe LV systolic dysfunction (FS <20%, fractional area change <25%) – Severe calcification or dysplasia of mitral leaflets – LA/RA thrombus or high thromboembolic risk	Contraindications due to procedural risk or lack of expected benefit.
Owner factors	Committed to advanced interventional therapy, understands risks/benefits, cost, and need for specialized follow−up (echo rechecks, anticoagulation).	TEER is not curative; it is palliative to improve QoL and survival.

- Reduced Regurgitation: The procedure typically leads to a 50-90% reduction in the severity of mitral regurgitation. Echocardiographic evaluations show significant decreases in the size of the left atrium and left ventricle.
- Improved Symptoms: Dogs often experience a rapid recovery and significant improvement in clinical signs, such as coughing and dyspnea, leading to a better quality of life. Many patients can be discharged within one to three days.
- Safety: The procedure has a relatively low risk of complications and a high survival-to-discharge rate (95% in one study).

Potential Risks and Complications

While less invasive than open-heart surgery, TEER is not without risks. The most serious and common complication is the detachment of the V-Clamp, which has a reported risk of approximately 7%. This can be fatal, with the highest risk period being the first 72 hours post-procedure. Other potential adverse events include leaflet tear, locking failure of the device, or device embolization. However, the risk of these events diminishes as healing occurs around the clamp in the weeks following the procedure.

References for Mitral V-clamp in dogs

1. *Sasaki K, Ma D, Mandour AS, Ozai Y, Yoshida T, Matsuura K, Takeuchi A, Cheng CJ, El-Husseiny HM, Hendawy H, Shimada K, Hamabe L, Uemura A, Tanaka R. Evaluation of Changes in the Cardiac Function before and after Transcatheter Edge-to-Edge Mitral Valve Repair in Healthy Dogs: Conventional and Novel Echocardiography. Animals (Basel). 2021 Dec 28;12(1):56.*
2. *Petchdee S, Pongkan W, Lei J, Jaturanratsamee K, Bootcha R, Meepoo W, Panprom C. Transcatheter Edge-to-Edge Repair of the Mitral Valve in Four Dogs: Preliminary Results Regarding Efficacy and Safety. Animals (Basel). 2024 Oct 24;14(21):3068.*
3. *Potter BM, Orton EC, Scansen BA, Abbott-Johnson KM, Visser LC, Chi IB, Ross ES, Del Nero B, Tantisuwat L, Krause ET, Rezende ML, Mama K. Clinical feasibility study of transcatheter edge-to-edge mitral valve repair in dogs with the canine V-Clamp device. Front Vet Sci. 2024 Dec 9;11:1448828.*
4. *Lee JM, Youp KA, Lee ST, Lee AR, Cho YW, Jung YS, Her J, Lee SK. Retrospective Evaluation of Transcatheter Edge-To-Edge Mitral Valve Repair in Dogs With Myxomatous Mitral Valve Disease. J Vet Intern Med. 2025 Jul-Aug;39(4):e70156.*

New Surgical Therapies

Mitral valve repair (MVR) is a highly specialized and technically demanding open-heart surgical procedure for dogs with advanced degenerative mitral valve disease (DMVD), also known as myxomatous mitral valve disease (MMVD). Unlike medical management, which is palliative, MVR is a definitive treatment that aims to correct the underlying cause of mitral regurgitation (the backflow of blood). The procedure is performed at a limited number of specialized veterinary centers worldwide.

Retrospective case series of transcatheter edge-to-edge mitral valve repair (TEER): V-clamp

Overview:

This retrospective case series evaluated the clinical characteristics and complications of **transcatheter edge-to-edge mitral valve repair (TEER)** in 20 dogs with MMVD. The study was one of the first to provide outcomes data for this emerging minimally invasive procedure in dogs.

Key Findings:

- **Procedure Success:** TEER was technically successful in all 20 dogs, achieving a significant reduction in the severity of mitral regurgitation.
- **Complications:** The complication rate was relatively low, with the most common being the temporary development of an abnormal rhythm.
- **Conclusion:** The study demonstrates that TEER is a feasible and safe procedure for dogs with MMVD. It offers a promising, less invasive option for managing advanced cases where conventional surgery may not be an option.

Reference:

*Lee, J.-M. et al. (2025). Retrospective Evaluation of Transcatheter Edge-To-Edge Mitral Valve Repair in Dogs With Myxomatous Mitral Valve Disease. **J Vet Intern Med**, 39, e70156.*

1. The Surgical Procedure

MVR is performed under general anesthesia with the dog on cardiopulmonary bypass (a heart-lung machine). This stops the heart temporarily, allowing the surgeon to operate on the valve in a bloodless, motionless field. The procedure primarily involves two key components to restore proper valve function:

- Artificial Chordal Replacement: In DMVD, the chordae tendineae (the "heartstrings" that tether the valve leaflets) often become elongated or rupture, causing the valve to prolapse and leak. The surgeon replaces these damaged chords with durable, artificial sutures, most commonly made of Gore-Tex. This restores proper tension and support to the valve leaflets.
- Mitral Annuloplasty: The mitral valve's ring-like structure (the annulus) can stretch and dilate due to the constant volume overload. This annular dilation further worsens the regurgitation. The surgeon performs an annuloplasty, typically by placing a purse-string suture around the valve annulus, to reduce its size and bring the valve leaflets closer together, thereby improving coaptation (their ability to close properly).

After the repair, the heart is restarted, the bypass is discontinued, and the heart's function is closely monitored as it takes over the circulation.

2. Patient Selection Criteria

Not all dogs with DMVD are candidates for MVR. The procedure is typically reserved for patients in the more advanced stages of the disease, as defined by the American College of Veterinary Internal Medicine (ACVIM) staging system. Ideal candidates usually fall into:

- ACVIM Stage B2: Dogs that are asymptomatic but have significant heart enlargement as a result of the mitral regurgitation.
- ACVIM Stage C: Dogs with a history of congestive heart failure (pulmonary edema or fluid in the lungs) that are currently managed with medication.
- ACVIM Stage D: Dogs with end-stage heart failure that is refractory to standard medical treatment.

Dogs with significant concurrent health issues, such as severe kidney or liver disease, coagulopathies, or those over 14 years of age, are generally considered poor candidates due to the increased risk of complications.

3. Outcomes, Prognosis, and Complications

The outcomes of MVR in carefully selected patients are generally very positive, with a significant improvement in both quality and length of life compared to medical management alone.

- Survival Rates: Published studies report high survival rates to hospital discharge, often exceeding 90% for dogs in Stage C and Stage B2. While the survival rate for Stage D dogs is lower, successful cases still show a marked improvement in their condition.
- Echocardiographic Improvements: The surgery leads to a significant reduction in heart size, with a decrease in the left atrial and left ventricular dimensions, which directly correlates to the reduction in mitral regurgitation.
- Clinical Benefits: Successful MVR often allows for the discontinuation of cardiac medications, resolves clinical signs of heart failure (e.g., coughing, breathing difficulty), and can lead to a prolonged, symptom-free life.
- Complications: While the procedure is highly successful, it carries inherent risks, including complications related to cardiopulmonary bypass and intensive post-operative care. These can include:
  - Thromboembolism (blood clots)
  - Acute renal failure
  - Myocardial dysfunction
  - Arrhythmias

MVR represents the most definitive treatment currently available for DMVD, offering a potential for a "cure" for the mechanical problem of the leaking valve, rather than just managing the symptoms. The procedure's success is heavily dependent on the skill and experience of the surgical and cardiology team, as well as meticulous patient selection and post-operative care.

Preoperative prediction models of mitral valve repair (MVR)

Overview:

This retrospective cohort study aimed to develop and evaluate preoperative prediction models for 30-day all-cause mortality following mitral valve repair (MVR) in dogs with myxomatous mitral valve disease (MMVD). The study used data from a single center to establish an objective risk assessment method for this novel surgical intervention.

Key Findings:

- **Predictive Model Development:** Two models were successfully developed: one for predicting all–cause mortality and another for cardiac death.
- **Significant Risk Factors:** Key factors identified as significant predictors of mortality included preoperative clinical signs of congestive heart failure, certain laboratory parameters, and echocardiographic measurements.
- **Conclusion:** The models provide a much–needed objective tool to predict the short–term prognosis of dogs undergoing MVR. This can help clinicians and owners make more informed decisions about pursuing surgical intervention and managing patient expectations.

Reference:

Furusato, S. et al. (2025). Preoperative Prediction Models for 30-Day All-Cause Mortality After Mitral Valve Repair in Dogs: A Single-Center Retrospective Cohort Study. ***J Vet Intern Med****, 39, e70152.*

Long-term outcomes of mitral valve repair (MVR)

Overview:

This study evaluated the long-term outcomes of mitral valve repair (MVR) using artificial chordae and annuloplasty in dogs with MMVD. The research aimed to identify risk factors and determine the overall prognosis for dogs undergoing this open-heart surgery.

Key Findings:

- **High Success Rate:** The long–term prognosis was favorable, with a high survival rate. The median survival time after surgery was significantly longer than expected with medical management alone.
- **Risk Factors:** The study identified several factors associated with a less favorable outcome, including preoperative presence of severe pulmonary hypertension and significant left atrial enlargement.
- **Conclusion:** MVR is a highly effective treatment for advanced MMVD, offering a good long–term prognosis for dogs who are suitable surgical candidates. The identification of key risk factors can help clinicians better select patients and manage expectations.

Reference:

Kurogochi, K. et al. (2025). Long-Term Outcomes of Mitral Valve Repair With Artificial Chordae and Annuloplasty for Myxomatous Mitral Valve Disease in Dogs. ***J Vet Intern Med****, 39, e70171.*

Imaging Diagnosis Guidelines for Myxomatous Mitral Valve Disease

Chest Radiographic Indicators

1. Chest Radiograph VHS Measurement

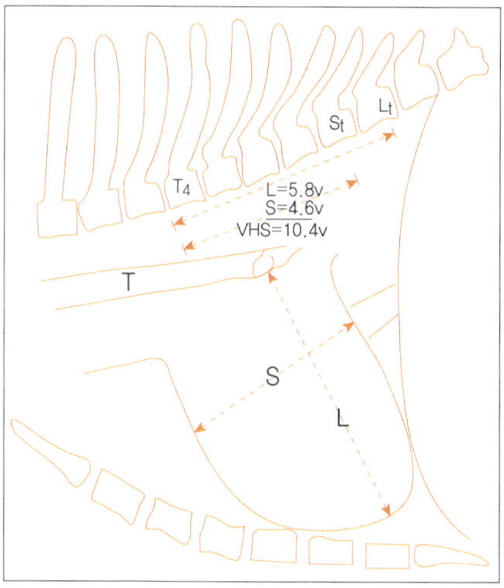

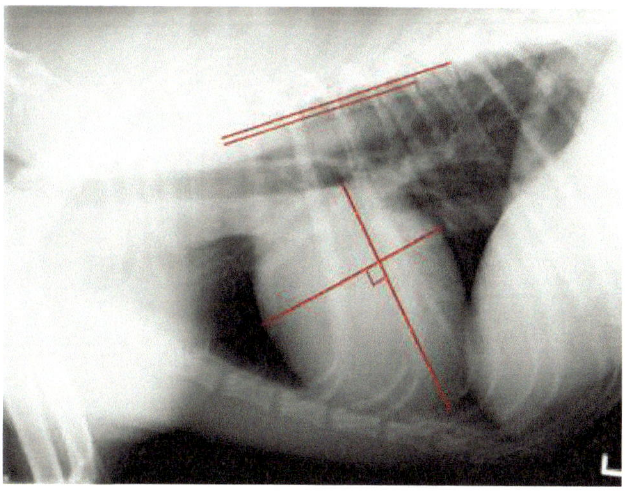

Vertebral heart score/vertebral heart size (VHS) for cardiac size measurement:

1. Take a right lateral thoracic radiograph.
2. Measure the distance from below the carina to the cardiac apex (straight line L). Mark this distance on a paper tape.
3. Measure the horizontal distance at the widest part of the heart, at a 90-degree angle to straight line L (straight line S). Mark this distance on the paper tape.
4. Place the starting points of straight line L (long axis) and straight line S (short axis) on the paper tape at the 4th thoracic vertebra (T4) position and mark the ending positions of straight line L and S on the radiograph.
5. Count the number of vertebral bodies covered by straight line L and S and sum the numbers. In the example provided, the long axis (L) covers 6 vertebrae and the short axis (S) covers 5 vertebrae. Thus, the VHS is 6+5=11.

Reference Values:

✓ **Boxer:** 11.6±0.8 v

✓ **Cavalier King Charles Spaniel:** 10.6±0.5 v

✓ **Dobermann:** 10.0±0.6 v

✓ **German Shepherd Dog:** 9.7±0.7 v

✓ **Labrador Retriever:** 10.8±0.6 v

✓ **Whippet:** 11.0±0.5 v

✓ **Yorkshire Terrier:** 9.7±0.5 v

✓ **Puppies:** 8.5~10.5 v

✓ **Cat:** 7.5±0.3 v (lateral view), cardiac width (short axis) in dorsoventral/ventrodorsal view is 3.4±0.25 v.

2. Vertebral Left Atrial Size (VLAS) Measurement

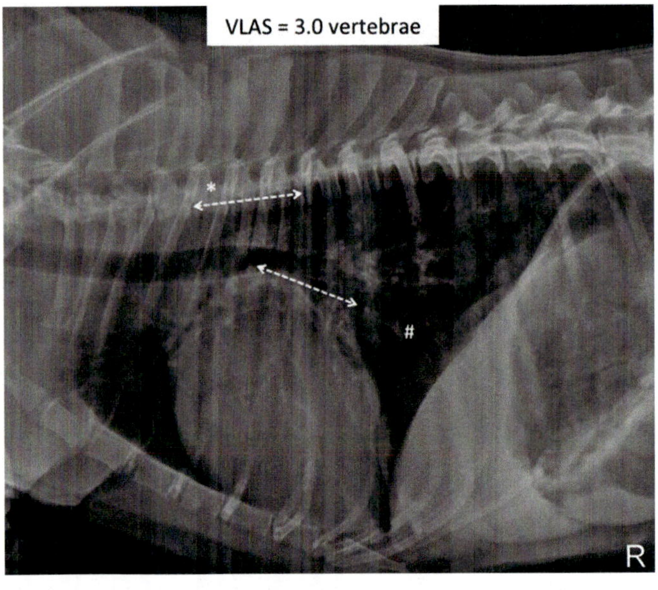

Vertebral Left Atrial Size (VLAS) Measurement Method: Draw a straight line from the ventral margin of the hilus to the point where the cranial margin of the T4 rib meets the dorsal margin of the caudal vena cava. Then, compare the number of thoracic vertebrae covered, similar to the VHS method.

Diagnosis Guide:

- **Control:** 2.1
- **ACVIM B1:** 2.1
- **ACVIM B2:** 2.6
- **ACVIM C-D:** 3.0

3. Radiographic Left Atrial Dimension (RLAD) Measurement

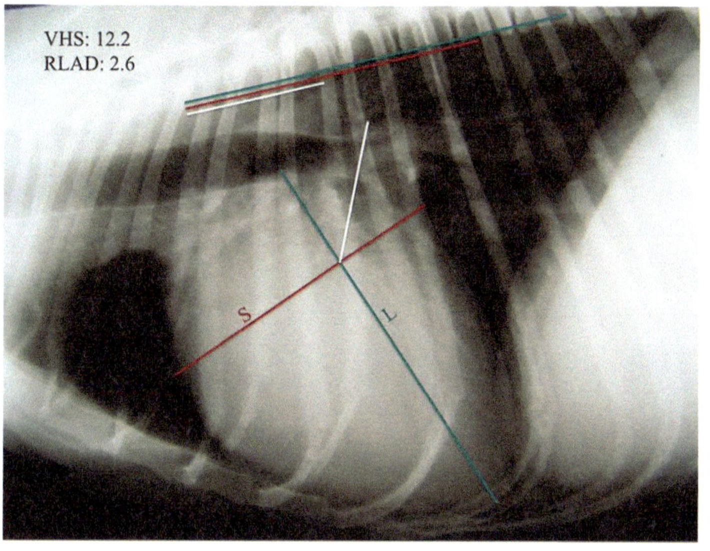

Radiographic Left Atrial Dimension (RLAD) Measurement Method: Take a right lateral thoracic radiograph. Draw a straight line from the ventral margin of the left main bronchus (hilus) to the intersection of the long and short axes measured in the VHS method. Then, compare the number of thoracic vertebrae covered, similar to the VHS method.

Diagnosis Guide:

- **RLAD > 1.8** indicates left atrial enlargement.

Key Echocardiographic Indicators Used by the Author

1. LA/Ao Measurement Method

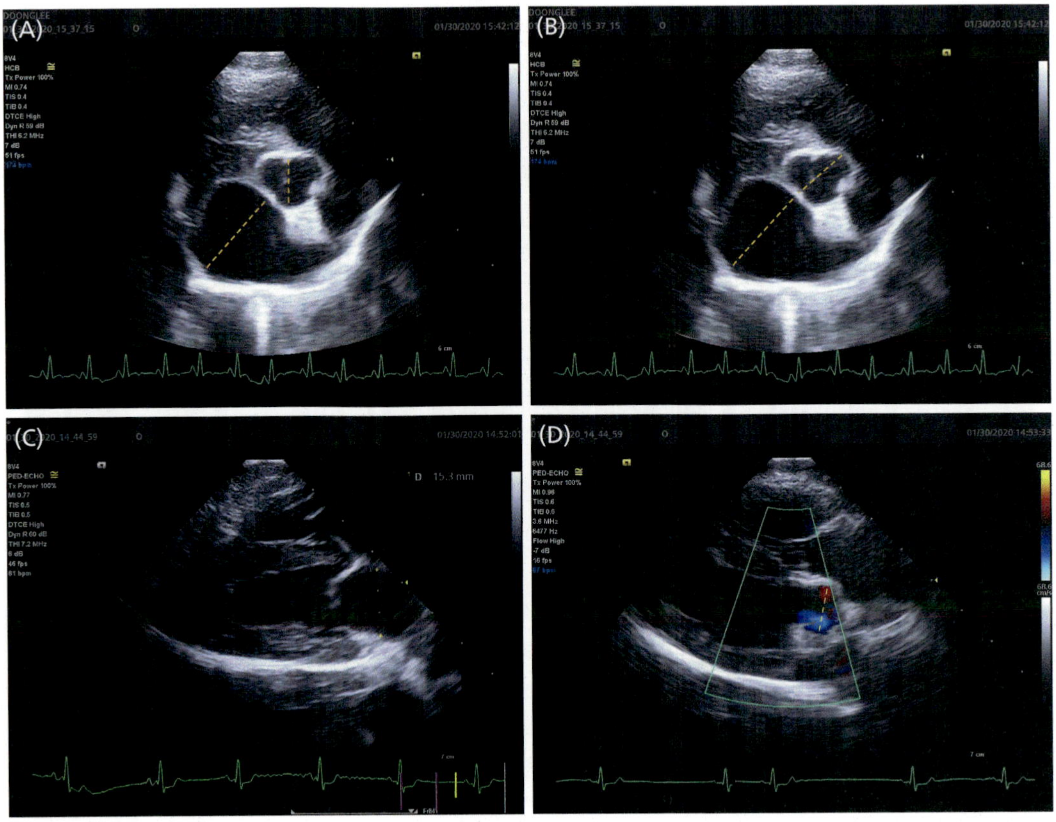

LA/Ao Diameter Measurement Method:

- **A: LA/Ao Sx (short axis) Measurement Method (Type 1):** In the short-axis view of the aorta, with the aortic valve closed, measure the inner diameter between the noncoronary aortic valve cusp and the right coronary aortic valve cusp. Measure the distance from the junction of the noncoronary aortic valve cusp and the left coronary aortic valve cusp to the distal end of the left atrium. Calculate the ratio of the two diameters.
- **B: LA/Ao Sx (short axis) Measurement Method (Type 2):** In the short-axis view of the aorta, with the aortic valve closed, measure the inner diameter between the noncoronary aortic valve cusp and the right coronary aortic valve cusp. Measure the distance from the junction of the noncoronary aortic valve cusp and the left coronary aortic valve cusp to the distal end of the left atrium. Calculate the ratio of the two diameters.

- **C & D: LA/Ao Lx (long axis) Measurement Method:** At the end of ventricular systole (end-systole: immediately after the T wave), just before the mitral valve opens, draw a vertical line from the middle of the interatrial septum to the opposite atrial wall, nearly horizontal to the mitral valve, and measure the diameter. When the aortic valve is open (early systole: at the start of the T wave), measure the widest diameter from the valve leaflet to the opposite aortic orifice. Calculate the ratio of the two diameters.

Diagnosis Guide:

- **LA/Ao Sx:**
  - **ISACHC I:** 1.4-1.8
  - **ISACHC II:** 1.8-2.0
  - **ISACHC III:** >2.0
- **LA/Ao Lx:**
  - **Normal:** 1.8-2.4
  - **LA/Ao Sx 1.6 = LA/Ao Lx 2.6**

2. LV Measurement Method

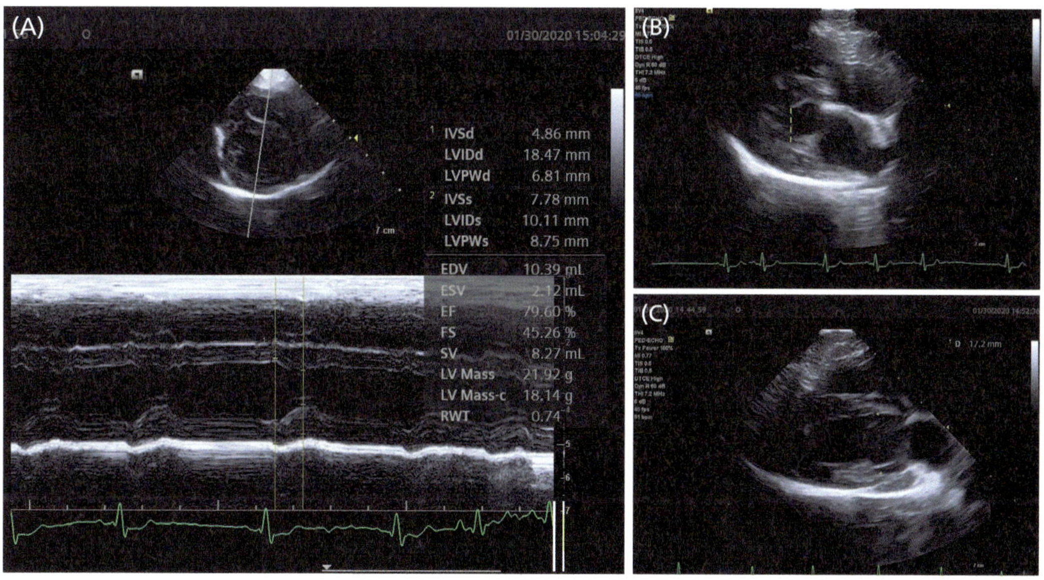

- **A. LVIDdN Measurement Method:** Measure the left ventricular end-diastolic dimension in the left ventricular short-axis view at the papillary muscle level. Calculate as LVIDd (cm) / body weight (kg)0.294.

- **B & C. LVIDd/Ao Ratio Measurement Method:** Just before the start of ventricular systole (at the very beginning of systole; immediately before the R wave), measure the widest diameter at the septal margin and the left ventricular free wall margin. When the aortic valve is open (early systole: at the start of the T wave), measure the widest diameter from the valve leaflet to the opposite aortic orifice. Calculate the ratio of the two diameters.

Diagnosis Guide:

- **LVIDdN > 1.7** indicates heart failure of ACVIM B2 stage or higher.
- **LVIDd/Ao:**
  - **ISACHC I:** <2.0
  - **ISACHC II:** 2.0-2.5
  - **ISACHC III:** >2.5

3. Assessment of MR Severity

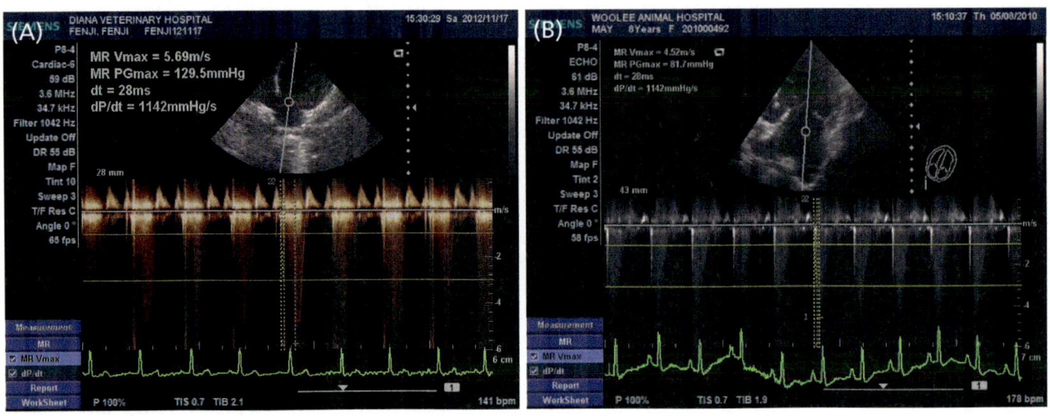

Assessment Method:

In the apical four-chamber view, measure MR using CW Doppler and then measure the slope of the MR flow file between 1 m and 3 m (dP/dt).

Diagnosis Guide:

- **> 1200 mmHg** indicates normal left ventricular systolic function.
- **< 800 mmHg** indicates systolic dysfunction.
- **Cut-off signs (arrows) as in Figure B** indicate severe (MMVD) Myxomatous mitral valve disease.

4. Transmitral E-peak and TDI E'-peak

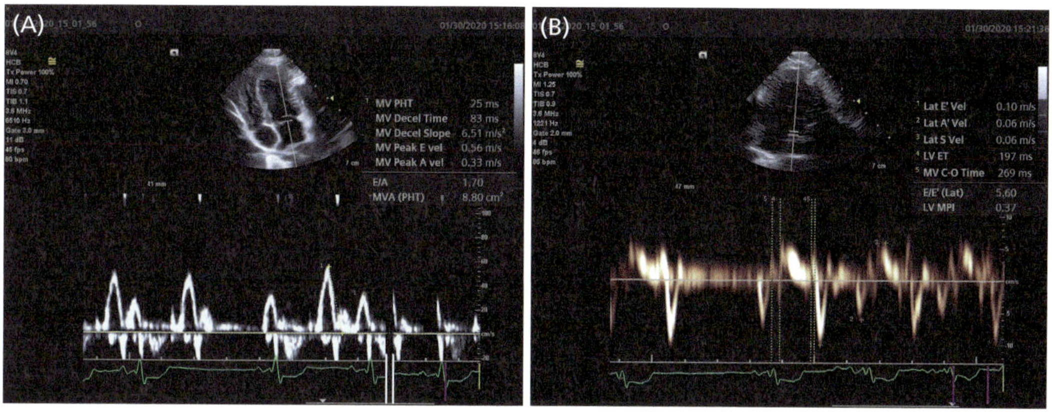

Measurement Method: In the apical four-chamber view, place the PW Doppler horizontally in front of the mitral annulus and measure the peak velocity of the mitral inflow. Switch to TDI mode and measure the peak velocity using PW Doppler at both the septum and the base of the mitral annulus (septal) and between the left ventricular free wall and the base of the mitral annulus (lateral). Calculate the E/E' ratio.

Diagnosis Guide:

- **E-peak:**
  - **ISACHC I:** < 0.8 m/s
  - **ISACHC II:** 0.8-1.2 m/s
  - **ISACHC III:** > 1.2 m/s
- **E/E' > 9** indicates LAP ≥20 mmHg.
- **E/E' ≥12** indicates congestive heart failure.

5. LV Tei index, E/IVRT, IVRT

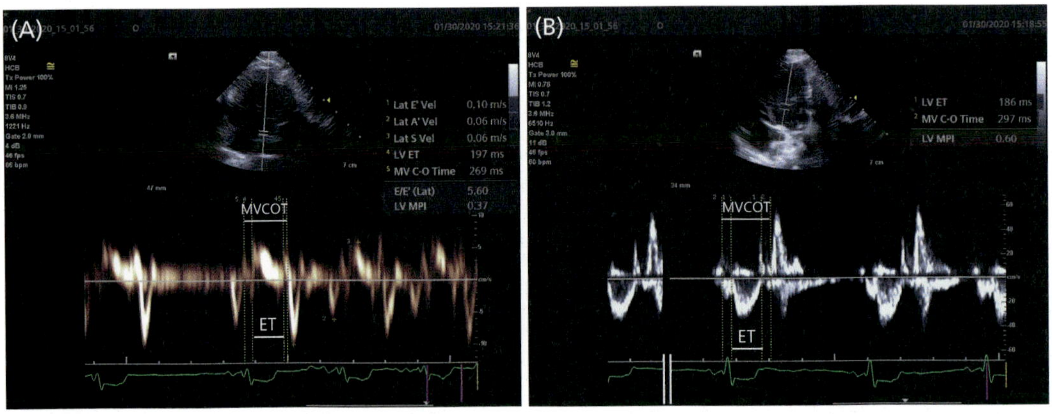

Measurement Method:

LV Tei index: In the apical four-chamber view, place the PW Doppler between the left ventricular outflow tract and the mitral annulus, then simultaneously measure aortic inflow and mitral inflow. On the ECG, the interval from the start to the end of the QRS complex is IVCT, and the interval from the end of aortic flow to the formation of the mitral E-peak is IVRT. Also, the time from the start to the end of aortic flow is LVET. The Tei index is (IVCT + IVRT) / LVET. **E/IVRT:** The value obtained by dividing the E-peak velocity of the mitral valve by the IVRT.

Diagnosis Guide:

- **E/IVRT ≥2.5** and **IVRT ≤45 msec** indicate the presence of congestive heart failure.

Index	Control	ISACHC Ia	ISACHC Ib	ISACHC II	ISACHC III
LV Tei	0.38±0.10	0.47±0.16	0.59±0.17	0.68±0.19	0.94±0.41
LVET (ms)	171±20	174±20	159±24	150±36	126±36

6. %LA-FS

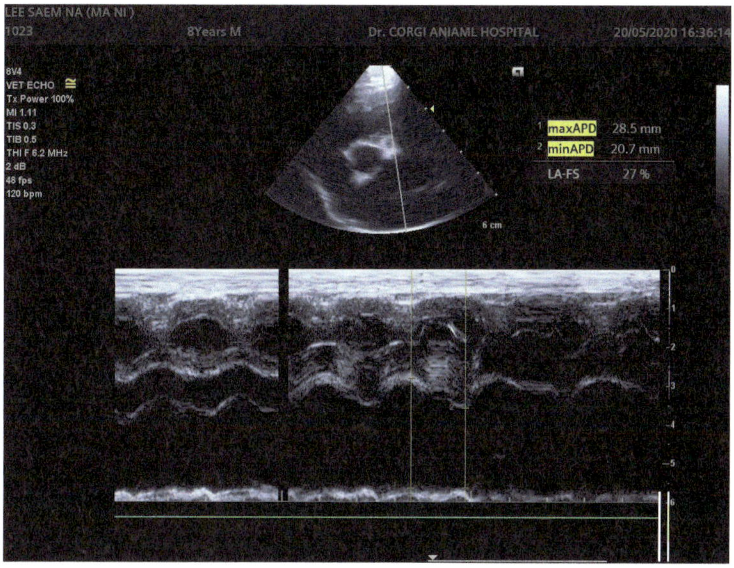

Measurement Method:

In the right parasternal long-axis view, place the M-mode cursor vertically through the center of the left atrium and measure the maximum and minimum diameters of the left atrium.

Diagnosis Guide:

- **Control:** 55.1±2.1%
- **ACVIM B:** 42.6±9.7%
- **ACVIM C & D:** < 20%

7. Echocardiographic Diagnostic Indicators for Pulmonary Hypertension

- **TR and PR velocity**

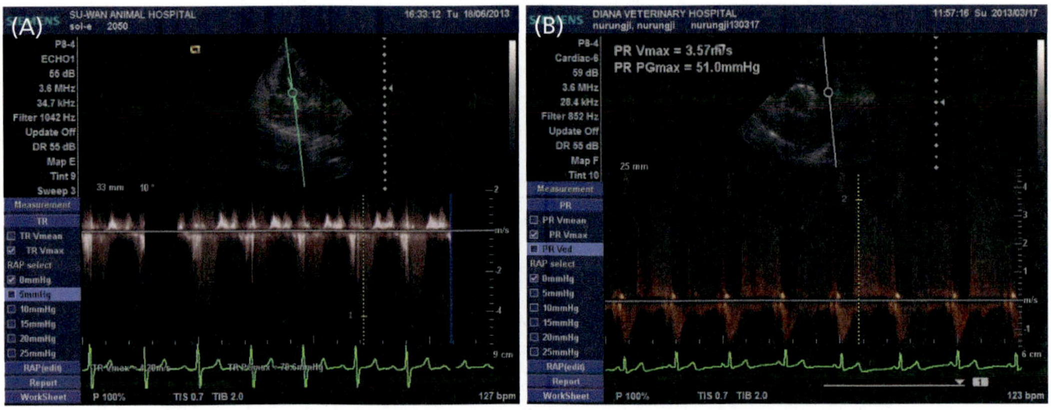

- **RPDAI**

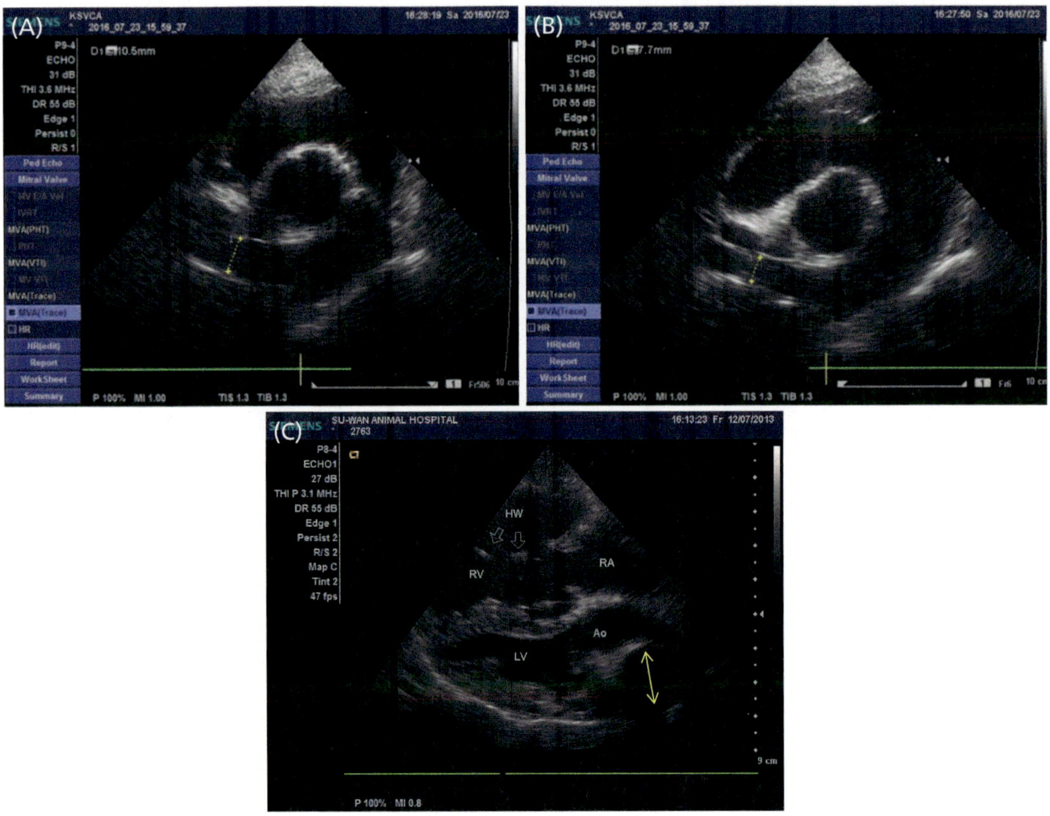

- **A: RVOT ET/AT & AT, B: Tricuspid E'/A', C: RVOT flow profile, D: RV Tei index**

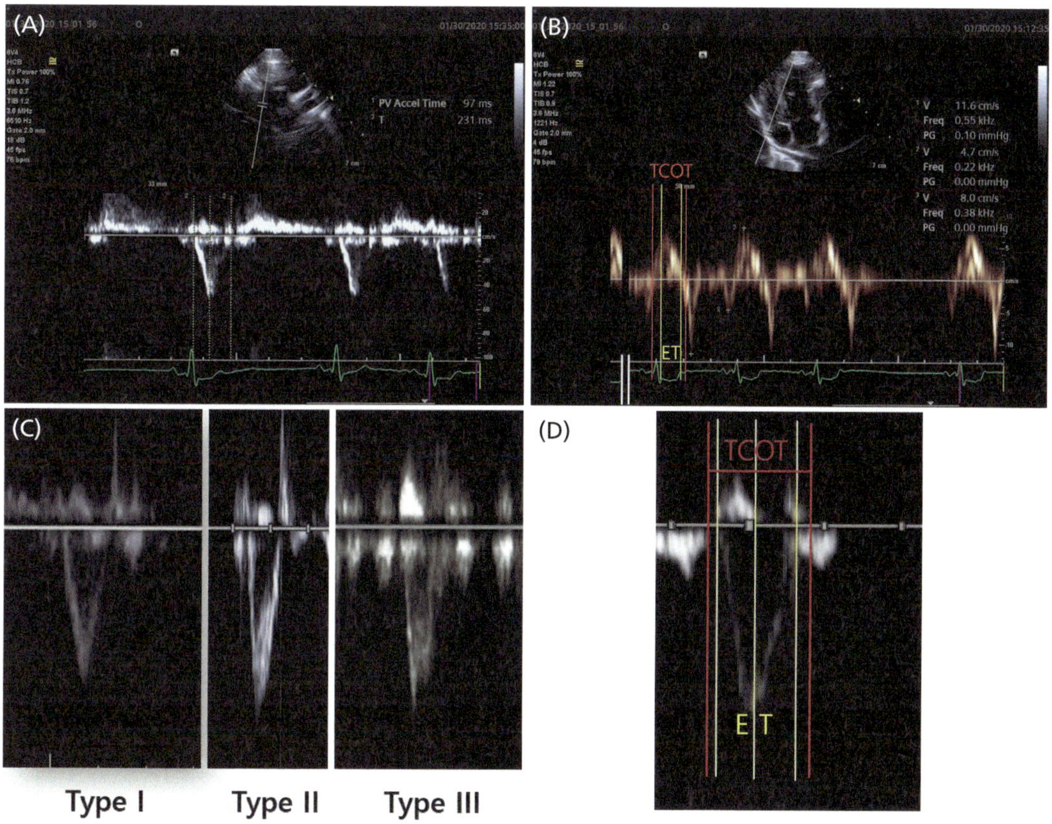

- **TAPSE**

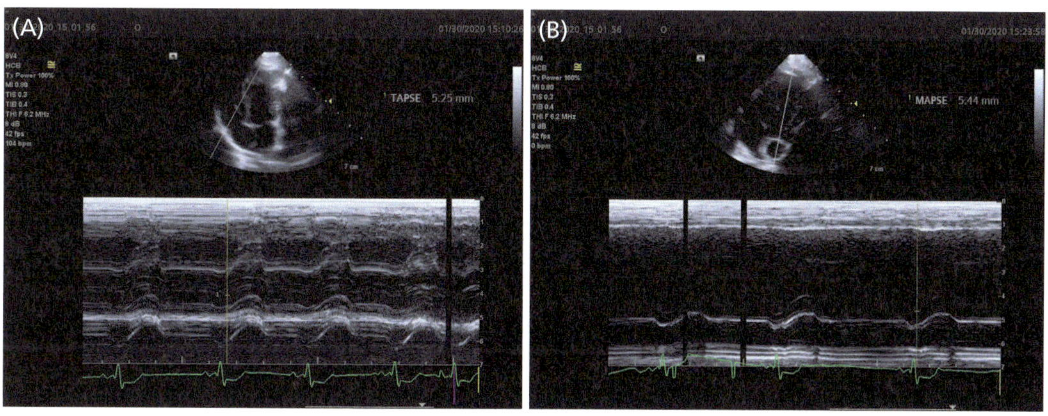

- **%RV-FS**

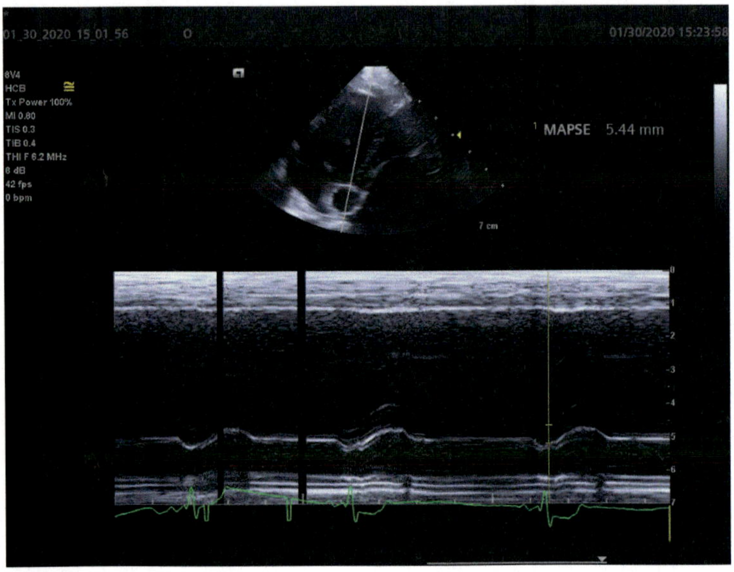

Diagnosis Guide:

Index	Mild	Moderate	Severe
TR peak velocity (m/s)	0–57	57–70	>70
PR peak velocity (m/s) mPAP	25–45	41–55	>55
RPDAI (PA)	20–31%	12–20%	<12%
RPDAI (LA)	28–35%	23–27%	<27%
AT/ET	0.28–0.41	0.28–0.26	0.24–0.26
AT (ms)	<52–58 ms implies PH		
RV Tei index	<0.28 no PH, >0.36 implies PH		
Global TDI	<11.8 implies PH		
Tricuspid E'/A'	<1.12 implies PH		
%RV–FS	>44% no PH		
nTAPSE (TAPSE/BW.kg$^{1/3}$)	ACVIM B1 0.60, B2 0.71, C 0.73		
TAPSE/Ao	<0.65 implies PH		

Index

A

K

L

M

N

| S |

| T |

U

V

X